TO THE OUTER PLANETS AND BEYOND

VOYAGER'S GRAND TOUR

HENRY C. DETHLOFF AND RONALD A. SCHORN

SMITHSONIAN BOOKS
Washington and London

© By special arrangement with the National Aeronautics and Space Administration, History Office, this publication is being offered for sale by Smithsonian Books, Washington, D.C. 20560-0950.

COPY EDITOR: Karin Kaufman
PRODUCTION EDITOR: Robert A. Poarch
DESIGNER: Janice Wheeler

Dethloff, Henry C.
Voyager's grand tour : to the outer planets and beyond / Henry C. Dethloff and Ronald A. Schorn.
p.   cm.
Includes bibliographical references and index.
ISBN 1-58834-124-0 (alk. paper)
1. Voyager project.   2. Outer planets.   3. Outer space—Exploration.
I. Schorn, Ronald A. (Ronald Anthony), 1935–   II. Title.
TL789.8.U6 V683 2003
919.9′204—dc21   2002029217

British Library Cataloguing-in-Publication Data is available

Manufactured in the United States of America
10 09 08 07 06 05 04 03   5 4 3 2 1

∞ The paper used in this publication meets the minimum requirements of the American National Standard for Information Sciences—Permanence of Paper for Printed Library Materials ANSI Z39.48-1984.

CONTENTS

Foreword
Edward C. Stone and Ellis D. Miner   v
Preface   vii
Acknowledgments   ix

Contents

FOREWORD

In 1977 two Voyager spacecraft were launched on unequaled journeys of discovery. No other mission has revealed so many diverse worlds close up for the first time and so vividly changed our view of the Solar System. Taking advantage of a rare alignment of the giant outer planets that occurs only every 176 years, Voyager 2 was propelled outward by the gravitational slingshot of successive planetary flybys, shortening to twelve years a normal thirty-year flight to Neptune. Had this alignment occurred five years earlier, our technology would have been too primitive for such a journey, and five years later, the United States had retired the capability to launch such a mission.

As the two spacecraft began their journeys, we knew there was much to be discovered, but we did not anticipate the wealth of discovery that lay ahead. Each day, as the Voyagers approached a planet, we saw what no one had seen before, discoveries that were superceded by even better ones as the spacecrafts' journeys continued. The trickle of discoveries rapidly became a flood that exceeded our ability to comprehend all that we were finding. It was both exhilarating and humbling.

The unexpected diversity of the Solar System contributed greatly to the success of the Voyager mission. Although we found many features familiar in form, they differed in detail. Dozens of hurricane-like storms riddle the Jovian atmosphere, and Jupiter's moon Io is one hundred times more active volcanically than Earth. The smooth, icy crust of Europa suggests there is an ocean beneath, and sulfur and oxygen shed by Io inflate Jupiter's magnetosphere into the largest object in the Solar System.

The dynamic nature of Saturn's rings was revealed by their rippled surfaces and by moons shepherding kinked, narrow rings, and organic molecules produced by the irradiation of the methane in Titan's deep atmosphere rain onto the moon's icy surface. The magnetic pole is unexpectedly near the equator on Uranus and Neptune, and there are cliffs 20 kilometers high on the complex surface of Uranus' small, icy moon Miranda. And even though Neptune's moon Triton is so cold that its nitrogen atmosphere is frozen into a polar cap, geysers erupt from its surface.

Of course, the scientific success was possible only because Voyager was an exceptional engineering success. The first fully autonomous interplanetary spacecraft, Voyager was designed for a four-year mission to Saturn. The capability to take and return images from Neptune, three times farther away, was developed only after successful Saturn flybys. The flight team reprogrammed the 8,192-word memories of the on-board computers controlling the spacecraft, and engineers of the Deep Space Network developed ways to combine many antennas on Earth to capture the much weaker signals from the outermost planets.

As this volume relates, Voyager was a bold undertaking for a newly spacefaring nation. Its success is a tribute to the innovation and dedication of many individuals in diverse organizations. For most involved, it was the journey of a lifetime, an unprecedented opportunity to share the excitement of discovery with people around the world.

The journey continues. The Voyager spacecraft are now in a race to find the edge of interstellar space while they still have enough electrical power to communicate back to Earth. They will become our first interstellar probes, carrying a message signifying that we have taken our first small steps into our local neighborhood in the Milky Way.

EDWARD C. STONE AND ELLIS D. MINER

PREFACE

The Voyager program is one of the most complex of all National Aeronautics and Space Administration (NASA) efforts. It is complex in its origins and conception, complex in its administrative history, and complex in its technical execution. Voyager 1 and 2 were planetary system explorers of extraordinary accomplishment, and the Voyager missions have dramatically transformed our knowledge of the outer planets and of the Solar System.

The Voyager Grand Tour is a product of creative management and a technology that developed its own dynamics, and it is in part the result of a unique culture and attitude that emanated from a small group of scientists and engineers at the California Institute of Technology's Jet Propulsion Laboratory. The Grand Tour is also a fruition of the effort mandated by Congress in 1958 in the National Aeronautics and Space Act to expand human knowledge of the atmosphere and space. The Voyagers are, in one sense, an extension of humanity's insatiable quest to know the unknown. The Voyager mission, and indeed, the technical capabilities of the spacecraft themselves, evolved even after launch.

Voyagers 1 and 2 were launched respectively in September and August 1977. At the time of launch, the missions were described as "Mariner-

Jupiter-Saturn, '77." By 1981, the Voyager spacecraft had completed scientific studies of Jupiter and the Jovian system, examined Saturn and its rings and moons, and were proceeding outward on a redefined mission to explore the outer planets, Uranus and Neptune. Then, having completed that work, and thus the preconceived Grand Tour of the outer planets, the Voyagers, in 1989, were now directed to extend their exploration beyond the Solar System on a new "Interstellar Mission." Subsequently, more than two decades after launch, in 1998, Voyager 1 passed Pioneer 10 to become the most distant artifact ever launched from Earth. The Voyager spacecraft remained "alive" and in contact with mission controllers and scientists some 6.5 billion miles distant on planet Earth. Even now, early in the twenty-first century, the Voyager exploratory vehicles are leaving the Solar System and probing the boundaries of true interstellar space.

We invite the readers to investigate with us this remarkable story on several levels. There is the story of the inception of the idea of planetary exploration, and then, how that idea began to achieve varying degrees of reality. The latter has to do with the story of Voyager engineering in a number of contexts. First, engineering in the context of designing and building a spacecraft that could endure decades of interplanetary flight and the then largely unknown hazards of outer space; second, engineering as it relates to the integration of science experiments into space-flight hardware; and third, engineering as it relates to flight engineering, trajectory design, spacecraft control, and communications. Voyager is in part the story of the integration of science and rocketry.

Finally, and most important, there is the story of Voyager science. How were Voyager science experiments selected? How have the Voyagers performed as instruments of planetary scientific exploration? What do we know today about Earth's space and planetary environment that was not known at the time of the Voyager launches in 1977? The author's have sought to carefully document Voyager discoveries. What might we expect to learn about interstellar space from the Voyager mission? Voyager is an ongoing mission.

To be sure, Voyager science and Voyager engineering, that is, Voyager history, is the work of countless numbers of men and women involved in the American, and indeed global, space and science community. Whatever else it may be, this too is their story and their history. The authors hope that this book will enable us to participate in that history and to come to understand more about our planetary environment and, indeed, about ourselves.

ACKNOWLEDGMENTS

We are indebted to many who gave of their time, consideration, and concern through the three years required to recreate the remarkable history of the Voyagers' Grand Tour of the outer planets. We would particularly like to acknowledge the contributions and support of Ellis D. Miner, deputy project scientist, who read and reviewed the work as it developed and who provided insightful advice and direction. Rex Ridenoure, a member of the Voyager Mission Planning Office team, helped organize and facilitate the extensive interview process, reviewed the manuscript, offered advice, and provided photographs from his personal collections. Charles Kohlhase, head of the Mission Planning Office and the Voyager "navigator," endured an extensive interview, read and corrected manuscript copy, provided photographs, and offered continuing counsel and encouragement.

The work could not have been completed without the professional and personal efforts of John F. Bluth, director of the Jet Propulsion Laboratory Archives and Records Center, and the archives staff. Bluth helped in ferreting out records and manuscripts and then read and critiqued preliminary draft chapters.

John Casani, Raymond L. Heacock, George Textor, Ronald F. Draper, and Ed B. Massey provided critical information and counsel through the interview process, as did Edward C. Stone Jr., who became Voyager project scientist in 1972 and continued to serve in that capacity after becoming director of the Caltech Jet Propulsion Laboratory in 1991. George Alexander, with JPL's Public Affairs Office, read and commented on early chapters of the history. Our work benefited greatly from the evaluations and input provided by representatives of the science and astronomy community, including Jay T. Bergstrahl, David Chandler, David Crawford, Gerard de Vaucouleurs, Stephen P. Maran, David Morrison, Bradford Smith, Arthur Lonne Lane, Garry Hunt, and others—and we are appreciative.

Finally, Dr. Roger Launius, NASA chief historian, provided the oversight, review, and direction to help keep us on course during the research and writing process, and Amy Paige Snyder, Steve Garber, and Douglas J. Mudgway offered a final review and critique. We wish to thank all who contributed in the production of this history. We hope that the reader will acquire a better understanding and share with us our wonder and appreciation for space exploration and our remarkable space and planetary environment.

1

FROM EARTH TO THE EDGE OF
THE UNIVERSE

At approximately 2:10 P.M. Pacific Time on February 17, 1998, Voyager 1, an interplanetary exploring vehicle located "in a dark, cold, vacant neighborhood near the very edge of our Solar System," became the most distant human-made object. Some 6.5 billion miles from Earth and traveling at a speed of 39,000 miles an hour, Voyager 1 outdistanced the older Pioneer 10 spacecraft, neared the edge of the Solar System, and pressed toward the invisible boundary that marks the beginning of interstellar space.[1] Voyager 1 and its twin, Voyager 2, in the course of their interplanetary travels, encountered and explored Jupiter, Saturn, Neptune, and Uranus, completing a remarkable "Grand Tour" of the giant outer planets that had begun almost a quarter of a century earlier. And that Grand Tour, at the beginning of the new millennium, became a new interstellar mission.

The Voyager program is one of the most complex of all National Aeronautics and Space Administration (NASA) efforts, complex in its origins and conception, complex in its administrative and political history, and complex in its technical execution. The two Voyager probes were plane-

1

tary-system explorers of extraordinary accomplishment, and the Voyager mission dramatically transformed our knowledge of both the outer planets and the Solar System.

The Voyager program was approved by NASA in 1972 after almost a decade of difficulties involving budgets, priorities, politics, public relations, and improvisations. An original Voyager mission to Mars, "a mission not flown," was specifically eliminated by Congress. A Grand Tour planetary exploration mission to the outer planets was conceived, scheduled, and then eliminated by budget cuts. Scientists and engineers at the California Institute of Technology's Jet Propulsion Laboratory (JPL) in Pasadena, California, proposed a less costly alternative program, Mariner/Jupiter/Saturn '77 (MJS '77), but there was much more in the MJS '77 bottle than indicated by that label. Eventually, the result of all this was the launch of the twin spacecraft from Cape Canaveral, Florida, in the summer of 1977, with a design life of five years and a narrowly defined mission of conducting closeup studies of Jupiter and Saturn, Saturn's rings, and the larger moons of those two planets.[2]

Voyager 1 took off for Saturn on September 5, 1977, two weeks after the launch of Voyager 2 on August 20. The former craft would actually reach Saturn first, since it pursued a faster trajectory, and NASA officials believed that designating it Voyager 1 would lessen public confusion. Voyager 1's key target was Saturn's outer moon, Titan. The course of Voyager 2, which would be trailing Voyager 1 by several months at the time of the projected November 1980 and August 1981 encounters with Saturn, would enable it to also visit Uranus and Neptune, but its primary mission was to back up Voyager 1. If Voyager 1 failed in its mission to Saturn and Titan, Voyager 2 was to be sent to Titan.[3]

Years after launch the mission was redefined. Engineers reconfigured Voyager 2's primary mission to the Solar System's two outermost giant planets and reprogrammed the flight systems to expand the probe's lifetime and capabilities. Indeed, a full-fledged Grand Tour resulted. The Voyager spacecraft approached Jupiter, the largest planet in the Solar System, in 1979. They investigated its largest satellites, Io, Europa, Ganymede, and Callisto, and discovered three new previously unknown moons: Adrastea, Metis, and Thebe. The twin probes visited Saturn in 1980 and 1981, examining that planet's winds and weather, its remarkable rings,

Launch of Voyager 1 on September 5, 1977. (NASA photo KSC-77PC-0296)

and its moons, ranging from distant Phoebe, believed to be a captured asteroid, to Titan, a planet-sized satellite with an atmosphere of nitrogen and methane that may once have resembled that of Earth.[4]

Voyager 2 made its closest approach to Uranus in January 1986, approximately 50,000 miles above the surface. The planet has a peculiar tilt to its axis (it "lies on its side") but has a magnetic field somewhat comparable to that of Earth. Voyager discovered ten new moons, bringing the planet's total to seventeen. Miranda, the smallest of the "classical" satellites known to have existed before Voyager 2's tour, was revealed as one of the strangest celestial bodies in the Solar System. More than three years later, on August 25, 1989, Voyager 2 flew within 3,000 miles of Neptune's highest clouds. At the time that planet was the most distant one from the Sun, because Pluto, in its highly elliptical orbit, was then closer to our Sun. Neptune orbits our star every 165 years and has the smallest diameter of the Solar System's gas giants but a larger mass than that of Uranus. Surprisingly, winds in Neptune's upper atmosphere can reach speeds of 1,200 miles per hour. Voyager discovered six of the eight moons that are now known to orbit the planet and closely investigated Triton, the largest satellite, which has a surface temperature of −391 degrees Fahrenheit, the coldest of any body in the Solar System.[5]

That done, Voyagers 1 and 2 had accomplished the Grand Tour of the outer planets. But there was more to come, for twenty years after launch both craft were still operational. Thus, in 1989 NASA reprogrammed them yet again. In their new interstellar mission, they would extend exploration of the Solar System beyond the outer planets and "search for the heliopause boundary, the outer limits of the Sun's magnetic field and the outward flow of the solar wind."[6]

At launch time, but even more so as the Voyagers began to make contact with the outer planets and previously unknown celestial bodies, their mission captured the imagination of the American people. The cast of *Star Trek,* a popular science fiction television series, made a full-length movie *(Star Trek: The Motion Picture)* centered on the discovery of an unexplained, interstellar vehicle named "V'ger" (pronounced "veeger"). A complex plot of intrigue and near-interplanetary warfare involving people of the Federation, the warlike and aggressive Klingons, and others, and an assault on Earth by an incredibly powerful "living" machine, is finally resolved by Captain Kirk and the crew of the starship *Enterprise* with the remarkable revelation that the central component of the machine is a vehicle launched from Earth in the year 1977 with the mission to "collect data and transmit it back to Earth." Through time and space the probe's identifying name "Voyager" had been corrupted to "V'ger."[7]

On a related plane, NASA public information releases compared the Voyager computers to "HAL," the system aboard the spaceship *Discovery* in Arthur C. Clarke's popular science fiction book (and later movie), *2001: A Space Odyssey.* The Voyager craft were "equipped with computer programming for autonomous fault protection," and their computers were described (at the time of launch) as "the most sophisticated ever designed for a deep-space probe."[8] These interplanetary explorers represented in some way the ultimate in a new age of machines, combining the American love of machines and fascination with computers. At a time when the American public was virtually falling in love with the personal computer (while retaining their old affection for the automobile), it was a combination that people could understand and appreciate.

The Voyager Grand Tour was and is a product of creative management and a technology that developed its own dynamics. It is in part the result of a unique culture and attitude that emanated from a small group of scientists and engineers—pioneers in rocketry and the advocacy of planetary exploration—at Caltech's JPL.

To be sure, the origins of the Grand Tour are complex and confused. They are in part one aspect of the effort mandated in 1958 by the National Aeronautics and Space Act to expand human knowledge of the atmosphere and space, and in part one facet of humanity's endless quest to know the unknown. In one sense the origin of what became the Voyager missions may be traced to the first time a human raised his hands to the heavens as though to touch the stars. One could argue that Nicholas Copernicus (1473–1543) provided the impetus for planetary exploration when he defined the Solar System with the Sun at its center and an Earth that revolved around that luminary along with the other planets, rather than all the heavenly bodies revolving about Earth. Similarly, the revolutionary studies of Johannes Kepler in the sixteenth century concluded that Earth and the other planets revolved in elliptical orbits about the Sun, and Galileo Galilei in the early seventeenth century made telescopic observations that began to impose a sense of scientific order on the universe. The discovery of the law of gravity and the laws of motion by Isaac Newton later in that century provided insight and understanding for planetary exploration. In spite of these advances, however, for the most part the heavenly bodies remained beyond the scope of modern scientific inquiry until the twentieth century.

Despite the fact that Congress in 1915 founded the National Advisory Committee for Aeronautics (NACA) "to supervise and direct the scientific study of the problems of flight with a view to their practical solution," voyages through space rather than our atmosphere remained largely in science fiction and fantasy's bailiwick. (Jules Verne's *De la Terre à la Lune [From the Earth to the Moon]* [1865] and his *Autour de la Lune [Around the Moon]* [1870], both of which remained popular well into the twentieth century, are notable examples.) Russia's Konstantin Eduardovich Tsiolkovsky (1857–1935) provided one spark that helped generate serious scientific and engineering interest in rocketry and interplanetary exploration. His "Investigation of Outer Space by Means of Reaction Apparatus," written in 1898 and published in 1903 by *Science Survey,* was not widely recognized or disseminated in Russia or, indeed, anywhere else until the 1923 publication in Germany of *Die Rakete zu den Planetenraumen (Rockets to the Planets)* by Hermann Oberth. In the 1920s Tsiolkovsky followed his earlier treatises with studies describing the development of multistage rockets, artificial earth satellites, and manned orbiting platforms as way stations for interplanetary flight. Subsequently, interest in rocketry and planetary exploration soared in Germany and the United States. German enthusiasts even organized the German Interplanetary Society.[9]

In the United States, Robert H. Goddard of Massachusetts, as had Tsiolkovsky in Russia, pursued a lonely, though in his case decidedly practical, interest in rocketry. He completed his thesis at Clark University, "A Method of Reaching Extreme Altitudes," in 1914, and after revisions published it in the *Smithsonian Miscellaneous Collections* in December 1919. His ideas received more skepticism and criticism than praise.[10]

However, among those intrigued with his work was George Ellery Hale, an astronomer and member of the Caltech Board of Trustees. As director of the Mount Wilson Observatory, Hale offered Goddard laboratory space for his rocket experiments. Under Caltech auspices, Goddard tested a small solid-fuel rocket for the Army Signal Corps in August 1918. Working independently, he constructed and launched the first successful liquid-propellant rocket at Auburn, Massachusetts, on March 16, 1926. Still failing to win recognition from his peers, and suspicious that his work was being plagiarized, Goddard left Clark University and moved his research to Roswell, New Mexico, where, with the help of Charles Lindbergh and a grant from the Daniel Guggenheim Fund for Aeronautics, he continued his work. By his own choice he isolated himself from the academic and scientific communities, and thus for many years his epic achievements were unrecognized and underappreciated.[11]

In the same year (1926) that Goddard began his experiments in rocketry in New Mexico, Caltech, headed by Robert A. Millikan, a Nobel Prize–winning physicist, secured funding from the Guggenheim Fund for Aeronautics to establish the Guggenheim Aeronautical Laboratory, California Institute of Technology (GALCIT). The laboratory constructed a wind tunnel and turned its talents to aerodynamic research and its aspirations to flight beyond Earth, inspired in part in the latter by its first director, Theodore von Kármán, a Hungarian-born graduate of the University of Berlin. His interests in rocketry and space flight were sustained by members of the GALCIT group, including John W. Parson, Ed Forman, Apollo M. O. Smith, Rudolf Schott, Hsue-shen Tsien, Weld Arnold, and Frank J. Malina, who collectively by 1939 had begun to focus on the design fundamentals of high-altitude rockets.[12]

In 1939 the National Academy of Sciences began collaborating with the U.S. Army Air Corps in basic research on the development of rocket motors, and subsequently, in 1940, the Air Corps awarded Caltech a contract to continue work on the design and development of both solid- and liquid-propellant rocket motors. That contract enabled the GALCIT group to build permanent test and laboratory facilities in the Arroyo Seco Canyon just north of Pasadena.[13] The liaison between the National Academy of Sciences and the Caltech science community infused JPL's early space flight aspirations with scientific initiatives.

Both space flight and space science, however, soon deferred to a new reality. In December 1941, with American entry into World War II, the military turned to the GALCIT Rocket Research Group for an accelerated program in rocket-motor development. GALCIT's initial successes, as well

JPL's GALCIT Rocket Research Group at their test site, c. 1939. (NASA photo P-9007A)

as pressures imposed by the deployment of Germany's V-2 military rockets, capable of hurling indiscriminate destruction hundreds of miles, contributed to expanded research in longer-range rocket motors, the development of missile guidance and control systems, and in 1944, the organization of Caltech's Jet Propulsion Laboratory. Theodore von Kármán became the JPL's first director and Frank J. Malina its chief engineer; both were committed to the idea that rockets could be used for interplanetary travel.[14] Theirs was a commitment and a dedication that pervaded the GALCIT/JPL community and contributed significantly, in time, to the reality of the Voyager Grand Tour.

Following initial successes with the laboratory's JATO-type (jet assisted takeoff) rocket motors for aircraft, JPL engineers designed the Private, a short-range (about 10 miles) unguided missile with a solid-propellant motor. Subsequently, work on a long-range, high-altitude sounding rocket resulted in the development of the WAC ("without attitude control") Corporal. Tests achieved increasingly higher altitudes until, in October 1945, shortly after the close of World War II, initial data indicated that a Corporal had reached a height of more than 40 miles and was thus (erroneously) thought to have been the first American-made object to "escape

Frank J. Malina. (JPL Archives photo M-7)

the Earth's atmosphere."[15] That launch, and subsequent Corporal successes, rejuvenated JPL's commitment to space exploration.

Within a month of the surrender of Japan, the U.S. Army asked Caltech to continue operating its Jet Propulsion Laboratory on a permanent basis in cooperation with special army "rocket units." Those units were composed of German V-2 rocket scientists and engineers who, at the close of the war in Europe, had elected to surrender to the United States rather than risk being taken prisoner by the Soviet army. Subsequently, JPL's rocket technology was fused with that of the Germans, producing longer-range ballistic missiles, including a WAC Corporal-Bumper vehicle launched in July 1950, from the White Sands Proving Grounds in New Mexico. The second stage rose 250 miles above Earth, far higher than any object had ever flown.[16]

That same year, the army relocated the German rocket group to the new Ordnance Guided Missile Center at Huntsville, Alabama, under the technical direction of Wernher von Braun. During the Korean War, 1950–53, and in the years immediately following, the U.S. Army, Air Force, and Navy conducted hundreds of launches of longer-range Redstone, Jupiter,

Juno, Atlas, and Polaris rockets, while JPL focused on electronics research, guidance and control, communications, and space science.

American and German rocketry achievements, however, were soon eclipsed by the Soviet Union's launch of Sputnik 1, the world's first artificial satellite, on October 4, 1957. Although a few in President Dwight D. Eisenhower's administration labeled Sputnik a "neat scientific trick" or a "silly bauble," the subsequent launch on November 3 of Sputnik 2, a much larger and heavier satellite carrying the dog Laika, precipitated great consternation and even alarm.[17] The Soviet satellites created a sense of crisis and led to significant changes in American education, reinvigorated American interests in math, science, medicine, and engineering, and had much to do with the establishment of an American space program and the subsequent exploration of the planets.

On November 25, 1957, the Preparedness Investigating Subcommittee of the Senate Committee on Armed Services, chaired by Texas senator Lyndon B. Johnson, began an intensive "Inquiry into Satellite and Missile Programs" that lasted through January 1958. Homer Joe Stewart, a member of the Caltech faculty and a systems analyst for the missile development program at JPL, served as a special consultant to the committee, along with William Houston, president of Rice University. In view of the "tremendous military and scientific achievement of Russia . . . our supremacy and even our equality has been challenged," Senator Johnson said. He suggested that Sputnik represented an even greater challenge than Pearl Harbor.[18] In response the United States immediately tried and failed to launch a U.S. Navy Vanguard rocket payload into orbit. Subsequently, the Army Ballistic Missile Agency (ABMA) was directed to launch a scientific satellite into orbit in celebration of the forthcoming International Geophysical Year.

The successful launch of America's first orbital satellite, the Army/JPL-designed Explorer 1 on January 31, 1958, using the Juno I (a modified, four-stage Jupiter C) booster, bolstered JPL's aspirations for planetary exploration. While the American public focused on the launch vehicles and on placing a scientific package in orbit about Earth, JPL engineers and scientists, particularly Eberhard Rechtin, JPL's chief of guidance research, recommended that technical efforts be concentrated on Mars, the Moon, and the other planets.[19] Martian and other planetary probes needed more powerful rockets than those that placed packages in Earth orbit, and, just as critically, sophisticated guidance, communications, and telemetry systems.

Rechtin urged the ABMA to join JPL in persuading the administration to commence a "visionary program of lunar and planetary missions" based upon the development of a launch vehicle (later defined as the Juno IV) that could deliver a 550-pound payload to the Moon or a 300-pound payload to Mars.

Concurrently, President Eisenhower's Scientific Advisory Committee (PSAC) and its special Space Science Panel advocated "contact of some type with the moon as soon as possible" as the proper response to the Soviet successes in space. Moreover, the contact should be of such significance that "the public can admire it."[20] Explorer 1, and subsequent American responses to Sputnik, gave the old GALCIT/JPL interest in space exploration new credibility.

Thus, on March 27, 1958, the administration authorized the Department of Defense to establish a scientific program to send five Pioneer probes, as they were called, to explore space in the vicinity of the Moon. The department's Advanced Research Project Agency (ARPA) managed the program, and the Air Force was given responsibility to develop and launch three lunar probes, while the army, in collaboration with JPL, was assigned responsibility for two.[21] The program helped hone JPL's and America's understanding of spacecraft design and eventually led to interplanetary Pioneers 6 to 12; Voyager 1 passed a Pioneer probe in deep space some 6.5 million miles and several decades distant from the inception of the program.

Acutely and uniquely aware of how important a "sophisticated instrumentation and communication capability" would be to any lunar or planetary program, Rechtin and his JPL colleagues—including Walter K. Victor, head of the electronics research section; Robertson Stevens, head of the guidance techniques research section; and William Merrick, head of the antenna structures and optics group—collaborated in the design of an electronic system that could economically and efficiently handle communications with lunar and planetary vehicles. In early February 1958, Rechtin on his own initiative, assigned Merrick's group the task of selecting an antenna design for space communications.[22]

The Jet Propulsion Laboratory's initiative in the early design and development of sophisticated, permanent communications antennae was a critical step in the exploration of space. In July 1958 JPL scientists submitted a comprehensive "Proposal for an Interplanetary Tracking Network" and actually began construction on their first antenna before receiving authorization of funding, again betraying JPL's propensity for independent thought and action. Scientists at JPL constructed an 85-foot-diameter, 960-MHZ antenna on a not-too-distant government-owned site at Goldstone Dry Lake in the Mojave Desert that had been used by JPL engineers to test rocket engines in the 1940s. The antenna was ready for service a month before the first Army/JPL Pioneer launch. The Goldstone facility became the cornerstone of what would become, over the next two decades, the Deep Space Network—essential to communications and control of Pioneer, Voyager, and other lunar and planetary spacecraft.[23] The ingredients necessary for space flight and space exploration were falling into place.

The essential component would be the National Aeronautics and Space Administration, established by Congress on July 29, 1958. The new organization, which inherited and replaced NACA and elements of other government organizations, was given a broadly defined mission that included "the expansion of human knowledge of phenomena in the atmosphere and space." It began organizing new space flight centers and soon incorporated the Jet Propulsion Laboratory. The terms of its contract with JPL, as with the contract between JPL and the army, gave the laboratory "wide discretion in its operations." JPL's pioneering work with the Private, WAC Corporal, Explorer 1, and its current research and development on the Pioneer program were all a part of the NASA inheritance. And there was more. Even as NASA was being organized, on November 7, 1958, JPL submitted a proposal to NASA administrator T. Keith Glennan for a space flight program that basically cast the laboratory in its self-perceived role as "NASA's major space-flight laboratory" or, as JPL director William Pickering explained, "the national space laboratory."[24]

With the establishment of NASA, the United States now had a civilian program for space exploration—but where would it go? Congress's Committee on Science and Astronautics established a Select Committee on Astronautics and Space Exploration to address that question even before

NASA's creation. The committee began hearings in March 1958 to consider "The Next Ten Years in Space, 1959–1969." Pickering was among the many who gave counsel. In his testimony he said that he anticipated, in time, the development of reliable chemical rocket engines having thrusts of several million pounds, prototypes of atomic rocket engines, the development of ion propulsion systems, and reliable solar power sources having an almost indefinite lifetime.[25]

Arthur C. Clarke, one of the world's leading authors of science and science fiction literature, as well as a physicist, astronomer, and mathematician, believed that one could expect "with a very high degree of assurance, almost amounting to certainty," automatic probes to the Moon, Mars, and Venus; the orbiting of meteorological and communications satellites; nuclear propulsion, and flights around the Moon, Mars, and Venus, with the landing of a spaceship on the Moon possible but unlikely.[26]

Communications satellites, "earth-orbital" vehicles, nuclear rockets, plasma or ion drives, astronomical observatories, space laboratories, space stations, and missions to the Moon, Venus, and Mars were all projected as very real possibilities within the next ten years. But there was, in all of this, no mention by those testifying before Congress, and little indication of concern, with the exploration of the outer planets of the Solar System. How then did Voyager, a project to explore those outer planets, come to be? It came about in part as the product of an evolving technology, for aerospace engineering was a learning experience.

The Pioneer probes, for example, experienced persistent failures in their early years. The first Pioneer, launched by the Air Force on August 17, 1958, used the Air Force Thor rocket for the first stage and Vanguard second- and third-stage rockets; it blew up after seventy-seven seconds when a turbine pump failed. NASA's Pioneer 1, launched on October 11, 1958, attained an altitude of 70,000-plus miles and thus failed to reach the Moon. When Pioneer 2 was launched on November 8, the third stage failed to ignite and separate from the second. Pioneer 3, launched on a Jupiter II booster with JPL-constructed solid-propellent upper stages, experienced guidance failures and was destroyed. The failure of these lunar probes was heightened when, on January 2, 1959, the Soviet Union launched Luna 1, the first space vehicle to escape Earth's gravitational pull. Luna 1 passed within 3,728 miles of the Moon's surface and then en-

Pioneer 4. (NASA photo 314-3944)

tered into orbit around the Sun. Unfortunately, the Luna 1 success diminished the luster of the Army/JPL Pioneer 4, launched successfully on March 3, 1959. This was the first American spacecraft to escape Earth's gravitational pull, but, unfortunately, it passed the Moon at too great a distance (37,200 miles) for its instruments to record any useful lunar data.[27]

Nevertheless, Pioneer 4's limited success offered an opportunity for JPL to press its agenda for interplanetary exploration as a way to regain American leadership in space. The laboratory presented NASA with a new program: Project Vega. There were to be lunar and Mars probes in 1960, a Venus flight in 1961, a meteorological Earth satellite, and two to four additional interplanetary flights to be determined later. On March 26, 1959, NASA accepted the program and awarded a contract for the Atlas first stage and related hardware to Convair Corporation on May 6.[28] Problems began immediately.

Although JPL was designated the project manager, NASA Headquarters provided oversight and approved contracts and progress payments to the contractors. The contractors, understandably, began to defer to Headquarters. In addition, the Vega program required close integration and cooperation among a number of the NASA centers, something contrary to JPL's experiences and culture and, indeed, something new to NASA. Within months of approving the Vega program, NASA rescheduled the first four planned missions for lunar flights and in December 1959 canceled it in favor of a system using the Atlas–Agena B combination instead of developing the proposed new upper-stage Vega rocket. It was "a bombshell," for JPL, Pickering said.[29] Nevertheless, the Vega program ended up having a relatively long-term impact on interplanetary exploration and ultimately on the Voyager Grand Tour.

In January 1959, with Vega authorization pending, NASA approved JPL's space communications network concept, which would better support an interplanetary program such as Vega, and entered into an agreement with the Department of Defense for the construction of tracking stations in Australia and South Africa. Those two stations, one located in a dry lake bed near Woomera, Australia, and the other located west of Pretoria, South Africa, along with the Goldstone installation in California and supplemental tracking stations scattered about the globe, comprised the Deep Space Network, which became a very important adjunct of JPL and of all U.S. lunar and planetary programs.[30] In addition, NASA's brief commitment to Vega helped reconfirm JPL's elected position as NASA's interplanetary "operations" laboratory.

Neither was JPL deterred in its interplanetary quest by the rising support from Congress and among the general public for putting humans into space. Engineers seized the moment to couple JPL's interest in planetary exploration with the developing human space-flight initiatives. In November 1959, just prior to the cancellation of the Vega program, a senior development engineer, Allan B. (Hap) Hazard, submitted "A Plan for Manned Lunar and Planetary Exploration" that included preliminary robotic scientific probes, limited lunar exploration by two-person teams beginning in 1966 or 1967, comprehensive surface exploration by ten- or twelve-person (specifically including men and women) teams using "moon mobiles" beginning in 1970, and the establishment of a permanent Moon base in the 1970s, followed by human reconnaissance and surface exploration of Mars by the mid-1970s. Primary launch vehicles would be the

1.2-million-pound-thrust Saturn rocket already under development, a proposed 5-million-pound-thrust Nova rocket, and, later, a 20-million-pound-thrust cluster of Nova rockets.[31]

Although disavowed as an "official" submission of the Jet Propulsion Laboratory, the proposal assumed that "manned Missions to Mars will occur first, and will be followed in some now uncertain order, by Missions to Mercury, Venus, and to the outer Planets and their satellites."[32] Thus, the Voyager Grand Tour, though still distant, had a conceptual framework as early as 1959.

The successful launch of Pioneer 5, NASA's first interplanetary mission, by the Air Force and the Space Technology Laboratory on March 11, 1960, and the placement of its scientific package (developed by the recently established Goddard Space Flight Center) into a solar orbit between Earth and Venus, helped move NASA into a new planetary program dimension. In response, it created the Planetary Science Programs Office, and Abe Silverstein, the former associate director of the NACA Lewis Flight Research Laboratory near Cleveland, Ohio, and now NASA's director of space flight programs, assigned John F. Clark to head that office. Clark then organized the Lunar and Planetary Programs Office under an assistant director, Edgar M. Cortright, who assumed budget and oversight authority for the Pioneer and subsequent planetary programs.[33]

NASA's focus in late 1959 and early 1960 was on the Moon. So was that of the Soviet Union. On September 12, 1959, the Soviets achieved another space spectacular by crashing their Luna 2 spacecraft onto the surface of the Moon. The following month, on October 2, Luna 3 orbited the Moon and photographed its far side. Although Pioneer 5 was diverted from a lunar orbit to an Earth-Venus orbit partly as a matter of "beating the Russians," the United States had not, as of spring 1960, matched Russian lunar accomplishments.

Using a vehicle design adapted from JPL's Vega planetary probe, NASA initiated the Ranger program for more aggressive lunar exploration. The Ranger was to be launched on an Atlas intercontinental ballistic missile topped with an Agena rocket, as opposed to the Thor-Able and Atlas-Able combinations used for many of the early Pioneers. Headquarters assigned JPL responsibility for the Ranger program, and James D. Burke, formerly JPL's deputy director of the Vega program, became program manager.

Harris M. "Bud" Schurmeier, chief of the JPL Systems Division, was responsible for flight trajectory design, systems integration, spacecraft assembly, pre-launch check-out, and flight operations. Although Ranger's objectives had to do with lunar exploration, its design, based on the Vega concept, related to interplanetary missions. The latter required unusual sophistication and what JPL's historian, Clayton Koppes, referred to as "engineering excellence," that is, the program required detailed design and fabrication and exceptional technical interfacing for the systems to work correctly.[34] Indeed, the technical design of the Ranger became a part of the emerging blueprint for the Voyager interstellar explorers. Moreover, many of JPL's future Voyager engineers and scientists had hands-on Ranger experience.

At NASA Headquarters, Ed Cortright selected Bill (Newton William) Cunningham to manage the Ranger program for the Office of Lunar and Planetary Programs at NASA Headquarters, and Oran Nicks was given overall liaison responsibilities with JPL. Nicks noted that Ranger was recognized as the first real spacecraft because it was "more than just something like the warhead you put on a missile . . . but in fact . . . a complex machine in its own right."[35]

Ranger provided a useful laboratory for learning to manage and mesh scientific objectives with engineering and cost realities, for science sometimes conflicted with technology. As Nicks explained it, there were "tremendous conflicts and trade offs which resulted from trying to develop the technology and do the science simultaneously."[36] Ranger moved NASA space technology one step closer to Voyager.

No sooner had Ranger come on line as a lunar project than, in May 1960, budget cuts, forced in part by the new priorities assigned lunar space flight, forced JPL to terminate more than one hundred employees.[37] Paradoxically, that same month, NASA approved another lunar program, Surveyor, to complement Ranger exploration. Surveyor was scheduled to soft land on the Moon and send back photographs and basic temperature, seismic, and geological data, and thus complement and support human lunar missions.

Headquarters assigned JPL responsibility for the Surveyor program. In addition, a new interplanetary project, called Mariner, was spun off of Ranger and assigned to JPL. Mariner came to be, in part, because NASA realized that planetary missions could be beneficially and inexpensively "piggybacked" on flight tests of the new Atlas-Centaur liquid hydrogen–oxygen rocket (acquired by NASA from the Army Research Projects Agency and under development by General Dynamics). Test flights would otherwise have to be flown with dummy payloads. Moreover, in early 1960 NASA had no ongoing programs for planetary exploration, and, finally, JPL "was aching to begin planetary missions."[38] Mariner was the direct technical progenitor of Voyager.

Robert J. Parks, a former student of JPL's director, Bill Pickering, who had managed the Sergeant missile program, then Ranger, became JPL's director of the Mariner (Centaur) program on June 30, 1960. In early July JPL submitted a proposal to NASA for a Mariner flyby mission to Venus in 1962 (Mariner A) and a soft-instrument landing on Mars in 1964 (Mariner B); NASA accepted the proposal on July 15. However, almost within months of receiving approval of the Mariner interplanetary program, NASA received information that progress on Centaur development lagged seriously. Not only was the program behind schedule, but the design itself was in question.[39] The news meant that at the least the scheduled launch of a Mariner probe to Venus for August 1962 was in jeopardy.

Nicks asked JPL's Dan Schneiderman, who had been a key engineer in the design of the Vega upper stage rocket, if an Agena rocket could be substituted for the Centaur upper stage. Although skeptical, Schneiderman investigated the problem and decided that it would indeed be possible to launch a twenty-pound payload on a planetary mission using the Atlas-Agena system.[40] Nicks next queried Silverstein about the possible use of an Agena rocket for the scheduled Venus launch in August 1962. Silverstein said nothing should preclude the use of the Agena in place of the Centaur. Nicks then called JPL to officially inform them of the Centaur problem and asked them to prepare a plan "for a substitute Venus mission using Ranger hardware." JPL "snapped at the opportunity," Nicks noted, and created a high-caliber team headed by Jack James (project manager) and Schneiderman (spacecraft systems manager) and had a proposal ready on August 28, 1961.[41] This marked the inception of the Mariner-R program, and was a formative moment in Voyager spacecraft technology.

Schneiderman, the spacecraft systems manager for the Mariner R program, traces Voyager's technical lineage to a common point of inception for Vega, Ranger, Mariner, and Mariner R. Bill Pickering, he said, in late 1958 or early 1959 (about the time of the December 1958 Pioneer 3 launch), asked him to design a 350-pound spacecraft to go to Mars. Out of that "came the Mariner and the Ranger." Schneiderman, Walter Downhower, Mark Comuntzis, John Casani, and Jim Burke worked up the conceptual idea and the specifications for a 350-pound Mars spacecraft.[42] This became the generic concept for future planetary probes, including Voyagers 1 and 2.

Next, a team of twenty engineers, scientists, and technicians, led by Schneiderman and including Casani, Eberhard Rechtin, and others, refined the concept and specifications and submitted a report on February 1, 1960, titled "Spacecraft Design Criteria and Considerations; General Concepts, Spacecraft S-1." That document became a textbook of the basic conceptual technology for the design and fabrication of spacecraft—including the Voyagers.[43]

The basic configuration included a hexagonal spaceframe (analogous to an aircraft's airframe) to which were attached antennas, solar panels, and packaged scientific components. The spaceframe was mounted on a truss which housed the navigational and electrical systems and separated the craft from the propulsion systems. The spacecraft carried a high-gain antenna composed of six equally spaced radial arms and a wire mesh reflector dish, solar panel assemblies, and accompanying (but varying) scientific packages, folded solar panels covered with photovoltaic (silicon) solar cells with backup power batteries, a telemetry system compatible with the Deep Space Network receivers, attitude-control gyroscopes, and cold-gas attitude-control systems.[44]

The maiden voyage of Mariner R in July 1962, unfortunately, fared no better than its Pioneer or Ranger predecessors. Mariner 1 was to fly by Venus, obtain data on the environment, and survey the planet's characteristics. Oran Nicks recalls arriving at the launch site at Cape Canaveral before dawn on the twenty-second: "Before launch the space vehicle was a breathtaking sight, poised and erect in the night sky, a great gleaming white projectile lit by searchlights so intense that their beams seemed like blue-white guywires." But within minutes of launch Mariner 1 "was trapped amid the flaming wreckage of an explosion that lit the night sky."

A range safety officer destroyed the Atlas-Agena as the launch vehicle began to veer sharply from its course. Dan Schneiderman called it the "world's greatest failure party."[45]

Mariner 2 fared much better. Built on site by JPL staff and subcontractors and launched the following month on August 27, Mariner 2 passed within 21,500 miles of Venus on December 14 and completed a forty-two-minute instrumental survey of the planet's atmosphere and surface. The last transmission from Mariner 2 was received on January 4, 1963, when it was 54 million miles from Earth. NASA, delighted that something "finally worked," in November 1962 approved the Mariner 3 and 4 flights based on an earlier JPL design study but substituting the Atlas–Agena D rockets for the Atlas-Centaur stack.[46]

Mariner 3, targeted for Mars and launched on November 5, 1964, failed to shed the shroud over its payload and was "lost in space" in orbit about the Sun. This was a bitter disappointment, Ron Draper recalled. Ronald F. Draper, born in Imperial, Nebraska, in 1933 came of age in Eureka, California, dreamed about space travel and life on Mars, became an Eagle Scout, and won a scholarship to Carnegie Tech, where he earned a degree in mechanical engineering with an aeronautics option. His first job was with Northrup Aviation working on the F89-J fighter and the SNARK missile. From there he went into military service, returned to Northrup in 1959, got his master's in aeronautical engineering at Caltech, and began work with JPL beginning June 19, 1961.[47]

Ted Pounder was his supervisor and Johnny Small managed the Systems Design Section; Dan Schneiderman was deputy director for the Mariner program. Work on Mariner 3 was already well along when Draper arrived, but he was there for the beginning of Mariner 4. It was a fantastic experience, he said—working seventy-two hours a week and wanting to do more. Mariner 4 was reconfigured on the basis of experiences with the Ranger and Mariner 2; the team improved the configuration control process, added new redundancies to prevent failures, and incorporated a traveling wave amplifier, among other things.[48] Each spacecraft benefited from the successes and failures of its predecessors.

John Casani, who had worked on both the Pioneer and Ranger programs and was the spacecraft systems engineer for Mariner 4, thought that the spacecraft had evolved and been refined so much that it had, in fact, become a new-generation spacecraft. Indeed, Voyager, he believed, was essentially a "leaner" Mariner 4, the first craft specifically designed to go into really deep space. It had more redundancy (that is, backup systems), a star tracker instead of an Earth-sensor, gimbal mechanisms to allow a camera to continually point in a desired direction, and a new zero-based digital memory system.[49]

Reared in Philadelphia, Casani entered the University of Pennsylvania to pursue a four-year program in classical studies but got deflected by his roommate into electrical engineering. Already a science fiction buff, in the year he graduated, 1955, he read an article in *Collier's* magazine by Wernher von Braun about the exploration of space. Casani was hooked. After an interview with Jack James at JPL, he hired on as an entry-level engineer to work on the Corporal missile guidance systems. He moved from the Corporal to become the payload systems engineer on the Pioneer lunar program, and from there to Ranger, Mariner, and eventually to Voyager.[50]

Mariner 4 incorporated new "architecture" and was a composite of the best of that which had gone before. As such, it represented the best of JPL's collective experience and technology. Launched on November 28, 1964, about three weeks after the failed launch of Mariner 3, Mariner 4 made a spectacular and error-free launch aboard an Atlas–Agena D rocket. Within two days it was half a million miles from Earth, drawing electrical power from its deployed solar panels and navigating with sensors locked on the Sun and the star Canopus. On December 5, a course change fired a 50-pound-thrust rocket motor, targeting Mariner 4 for an encounter from behind Mars and north of its equator on July 15, 1965. It arrived one day earlier than scheduled. On July 14 Mariner passed within 6,100 miles of the Martian surface, sending back the first close-range images of Mars and revealing the existence of impact craters on its surface.[51] The world's ancient fascination with Mars was stirred, and old speculations about canals, water, and Martians appeared to be permanently laid to rest.

Even before launch, Mars had brought the sometimes antagonistic Caltech campus and JPL "rocket scientists," along with the broader scientific community, into a common cause and a new excitement and anticipation. The Space Science Board of the National Academy of Sciences strongly advised in October 1964 that a program emphasizing the resolution of the question of life on Mars and a general scientific exploration of the planet be a preeminent "National Goal in Space." In fact, Caltech president Lee A. DuBridge and JPL director Pickering had already agreed to establish a Campus-JPL Mars exploration study group. On the day of the Mariner 4 launch DuBridge asked Bruce Murray, a Caltech geologist and strong advocate of scientific planetary exploration, to head a new Mars Study Committee. The committee adopted what might best be described as an "intense" agenda to evaluate Mars mission science objectives, priorities, scheduling, scale or size of the effort (Saturn vs. Atlas-Centaur), and degree of complexity, as well as to evaluate NASA-Caltech-JPL relations and responsibilities.[52]

The result of the concentrated dialogue between the Caltech campus, JPL, and NASA Headquarters was a Caltech report dated February 23, 1965, entitled "Suggestions for Martian Exploration Following Mariner IV." The primary objective given in that document for an orbiter-lander exploration of Mars was to determine the presence (or absence) of life on the Red Planet, and the nature of any such life, and to obtain general scientific knowledge of Mars. Exploration was to be initiated with flybys launched with Atlas-Centaur vehicles, followed by larger Voyager landers carrying photographic, scientific reconnaissance, and life-detection experiments, all supported by ground-based observations. The Voyager name was salvaged from JPL's earlier (spring of 1960) proposed Venus and Mars probes that were to be launched on Atlas-Vega boosters. In 1965 NASA approved the first Mars Voyager budget and established the initial objective: to determine the presence of life on Mars. In July, NASA assigned the Jet Propulsion Laboratory responsibility for managing development of the Voyager landing capsule.[53]

Meanwhile, NASA and media attention focused for the most part on Project Gemini, a two-person, maneuverable spacecraft, and on the development of the Apollo-Saturn lunar program. Three successful flights to the Moon in early 1965 (Rangers 7, 8, and 9) returned photographs of the surface before the craft made impact but went relatively unnoticed in the popular press as attention focused on manned programs. The first piloted Gemini flight, on March 23, carried Virgil I. Grissom and John W. Young through three orbits around Earth and a manually controlled reentry. Gemini 4, launched aboard a Titan II rocket on June 3, with James A.

McDivitt and Edward H. White, established another American space feat when White completed a twenty-two-minute "walk in space" (EVA, or extra vehicular activity); the spacecraft returned to Earth on June 7. Next, Gordon Cooper and Charles Conrad Jr. flew Gemini 5 through a week of orbital flight and simulated rendezvous maneuvers in August, while Gemini 6, with Walter M. Schirra Jr. and Thomas P. Stafford, completed the first rendezvous in space, joining the Gemini 7 spacecraft, which had aboard astronauts Frank Borman and James A. Lovell Jr.[54] The spectacularly successful Gemini program, which had three additional launches scheduled for 1966, provided the technological and operational knowledge for the Saturn-Apollo lunar program and lessened somewhat the urgency of the Mars Voyager program. Costs and other considerations began to cause "slippages," modifications, and, soon, the demise of the Mars Voyager program.

Federal expenditures increased 100 percent in the decade between 1960 and 1970, much of the new government spending going to defense, space, health, social welfare programs, and education. American military commitments in South Vietnam rose precipitously as the military situation and the general health of the domestic economy deteriorated. At home, war protests, civil rights marches, campus dissent, domestic violence, and various shades and professions of "liberalism" and "conservatism" created constant chaos and confusion. In 1965 and 1966, just as military actions in South Vietnam escalated into a full-scale, expensive, and controversial war, and the full measure of financial obligations to public education, urban renewal, and social welfare became due, the American space program in general, and the Apollo lunar program in particular, began to demand increasingly larger federal allocations. Expenditures on the space program which had totaled only $750 million in 1961, by 1965 exceeded $5 billion. There was reason to be concerned about the future of any costly program of space exploration.

The "space race" became increasingly competitive and increasingly expensive. Soviet scientists soft-landed a Russian spacecraft on the Moon on January 31, 1966. Luna 9 returned photographs of the lunar surface to Earth for three days. Then, in March, the Soviet's Luna 10 entered a lunar orbit, circling the Moon for ninety days. Finally, following an extremely intense flight readiness review, on May 30, 1966, NASA's Surveyor 1 lifted off, made the flight to the Moon in about sixty-four hours, soft landed, and began returning the first of more than eleven thousand photographs of the lunar surface. The previous month, NASA dedicated JPL's Deep Space Network's new, larger antenna at Goldstone located in the Mojave Desert near Barstow, California. Built under the management of Rohr Corporation in San Diego, this 210-foot parabolic aluminum dish, as tall as a twenty-one-story building and weighing more than eight thousand tons, more than quadrupled the collecting area of the largest U.S. 85-foot antenna then in service and provided deep-space tracking, communications, and sensitivity not available anywhere else in the world (although a British dish at the Jodrell Bank Observatory was larger by 40 feet). The enhanced capabilities of the antenna, for example, extended the useful life of the Pioneer spacecraft then in orbit about the Sun for an additional year. As carefully delineated by Douglas J. Mudgway in *Uplink-Downlink: A History of the Deep Space Network, 1957–1997* (2001) the Deep Space Network made Voyager's deep-space missions possible.[55] But despite the challenges, and the successes, the American space program became increasingly cost-driven.

By delaying Mars Voyager test launches and substituting the more powerful Saturn 5 launch vehicle for the Saturn 1B–Centaur, NASA created temporary "savings" for the Voyager program and used those savings to fund two Mariner Venus probes for 1967, two Mariner Mars probes for 1969, and two Mariner orbital missions to Mars for 1971, the latter to fly in conjunction with anticipated Voyager landings.[56] The shift back to Mariner, however, made the Voyager Mars missions even less certain.

Then, Congressman Joseph E. Karth and members of House Space Science and Applications Subcommittee of the Science and Astronautics Committee (which had oversight for NASA budgets) filed a very critical review of the Surveyor program being managed by JPL, noting planning, design, technical, and management deficiencies. Later, in the spring of 1966, Karth led a group in Congress determined to impose greater congressional control on the American space program. Subsequently, the House Space Committee sliced NASA requests for the Mars Voyager program from $71.5 million for 1968 to $50 million, and a Senate committee eliminated funding for the program completely. A congressional conference produced a $43 million budget compromise, when, on August 18,

1967, in fear of escalating costs, the House of Representatives halted all funding for the Voyager program. Representative J. Edward Roush, a member of the House Space Committee, explained that although the Mars Voyager program was valuable and desirable, it, like other programs, was a casualty of the Vietnam War.[57]

JPL suspended all Voyager-related work at the laboratory in November. For the record, John E. Naugle, the associate administrator for the Office of Space Science and Applications, issued a terse, so-called Project Approval Document, which "documented" termination of the Voyager program.[58] Mars Voyager was seemingly dead, but the mission would be reformatted as Project Viking, which sent two orbiter-landers to the Red Planet in the 1970s. And the Mars Voyager survived in other ways: the technology it represented, the engineering and scientific talent it reflected, and even the name—Voyager—endured. Mars Voyager had created a new consensus in the JPL-Caltech community and a higher level of determination. The human initiative and spirit that drove Voyager remained very much alive. As usual, NASA and JPL responded to the termination with innovative alternatives. First were "extended" Mariner missions, followed by a proposal for a Grand Tour of the outer planets.

A GRAND TOUR OF THE
OUTER PLANETS

Upon the elimination of the Mars Voyager program, NASA turned again to Mariner, a planetary explorer with lower costs and proven capabilities, scheduling Mariners 6 and 7 for orbital exploration of Mars in 1969. In addition, as a response to JPL's long interest in planetary missions, and the growing interest of the science community in such exploration, NASA asked JPL to establish the Advanced Planetary Missions Technology program designed to nurture the research and technology that would be necessary for extended planetary missions. The "contemplated missions would involve flights to the planets Venus, Mars, Mercury, and Jupiter" and might utilize Atlas-Centaur-, Titan III–, and Saturn V–type launch vehicles. The basic idea was to capitalize on the technological and scientific advances resulting from the Ranger, Surveyor, and Mariner projects.[1] One should also note that a key element in the Voyager Grand Tour of the outer planets was simply the realization that it could be done.

Sending a probe of any kind to the outer planets presents severe problems. Even in the course of the Mariner, Surveyor, and Pioneer programs, when it became clear that it was practical to explore directly "inner" planets such as Venus and Mars, it was clear that such tasks would be infinitely more difficult in the cases of more distant bodies such as Jupiter and Saturn. The difficulties would be further compounded for even more remote objects of the Solar System, such as Uranus, Neptune, and Pluto.

Given the great expanses of time and distance, along with the state of rocket technology, just getting to the outer planets seemed to be an insurmountable problem. The most efficient direct trajectory from Earth to another planet, that is, the one requiring the least energy, is the segment of an elliptical orbit whose closest distance from the Sun is our planet's orbit and whose farthest distance is the orbit of the target body. This trajectory, the "Hohmann Transfer," was named for Walter Hohmann, a German architect who supplemented Hermann Oberth's *Die Rakete zu den Planetenräumen (The Rocket into Interplanetary Space)* published in 1923, with his own *Die Erreichbarkeit der Himmelskörper (The Possibility of Reaching Celestial Bodies)* in 1925. Hohmann completed the first detailed calculations of interplanetary orbits.

Unfortunately, to reach Neptune on a Hohmann trajectory using the most powerful U.S. rocket in the 1960s, the Saturn V, would take thirty years! It is difficult to imagine a mechanical device, with a self-contained power source, operating flawlessly for thirty years without maintenance. It is equally difficult to imagine a society that would provide the funding and institutional support required for such a voyage in search of, for lack of better term, the unknown. It is also hard to visualize the scientists and engineers who can individually sustain such a long-term commitment to a project that could only achieve a degree of success—assuming the probe reached its target—after thirty years; many, if not most, of those who launched the mission would not be there at its close. And if the probe did reach its objective three decades after launch, would the results, the new knowledge obtained, justify the mission? All of these considerations were elements in deciding whether to explore the outer planets.

Even given a thirty-year mission over the most favorable trajectory, the booster rocket that launched such a mission would have to provide such a high initial velocity that only an enormous and costly vehicle, such as a Saturn V, would be able to loft a useful payload. In the late 1950s and early 1960s many searched for solutions to the seemingly insurmountable problems of the times and distances involved in space travel. There were proposals to use exotic power plants such as nuclear rockets, ion (electric) drives, and so-called solar sails, but such devices lay far in the future, and some are not yet, and may never be, practical.

Remarkably, there was a solution to the problem. It had been around for literally centuries and involved no new science or engineering technology. Isaac Newton laid the modern foundation of what came to be known as "celestial mechanics" (how objects in the Solar System move) in his 1687 treatise *Philosophiae naturalis principia mathematica.* In that work he solved completely and exactly the problem of how two celestial bodies move under the influence of their mutual gravitational attraction. However, when three or more bodies are involved things get very much more complicated; neither Newton nor scientists and mathematicians since have found a useful general solution even to the "three-body problem." This study was of great practical importance in Newton's time, for the position of our Moon with respect to the background stars held out the promise of being able to determine longitude at sea. Unfortunately, in the lunar case the gravitational effects of Earth, Moon, and Sun all had to be taken into account, and these combined to defeat even Newton.

Though no general solutions appeared that would define the interrelationships of more than two bodies, celestial mechanicians over the years discovered some interesting properties of particular solutions. This was especially true for the "restricted" three-body problem, where one of the objects involved has a mass so small that its gravitational effects on the other two can be ignored—but not their effects on it! In particular, three or more bodies can transfer energy among themselves, including energy of motion.[2]

A spectacular example of such behavior was provided by the curious history of Comet Lexell, discovered in 1770. It came relatively close to Earth (just over a million miles), became visible to the unaided eye, and turned out to have an orbital period around the Sun of only 5.6 years. Why had no one seen it before? The answer was provided by Anders Johan Lexell (after whom the comet was eventually named), who pointed out that in 1767 the comet had approached close enough to Jupiter that the former's orbit was changed drastically. Before that it presumably followed an orbit that never came close enough to the Sun to make it visible to observers on Earth. (As an aside, in its altered orbit, the comet passed very close to our planet as comets go and thus was, in a sense, an early example of a "Jupiter probe in reverse.") The comet was not observed at its next apparition in 1776 because it was unfavorably placed with respect to the Sun as seen from Earth. Lexell noted that the comet would have another close approach to Jupiter in 1779 that might change its orbit again and, indeed, in 1781 and 1782 astronomers looked for the comet, but in vain. The reason for the non-appearance was demonstrated convincingly by Urbain Jean-Joseph Leverrier in 1857; what Jupiter had once given, Jupiter had later taken away. The gravitational force of Jupiter had altered the comet's orbit in 1767, bringing it closer to Earth in 1770; and altered it again in 1779. There has been no further trace of Comet Lexell to this day. Evidently its orbit has been altered so much that it no longer approaches the Sun or Earth closely enough to be observed.[3] Significantly, Comet Lexell changed orbits without the aid of rocket motors.

It is important to understand that no energy is being created in such a case. Instead, energy (and in particular energy of motion, or kinetic energy) is strictly conserved. However, the effects of the transfer of energy on the motion of Comet Lexell, with its minuscule mass, were dramatic, whereas the resultant change in the motion of giant Jupiter, not to mention that of the Sun, was insignificant.

A simple terrestrial analogy of such an exchange of energy has long been used by E. Miles Standish, an astronomer at JPL. As he explains, just sit next to a railroad track and watch a passenger train go by. Now imagine someone walking down the aisle toward the rear of the train. His or her kinetic energy depends on velocity and, as the observer sees it, the pedestrian's motion is slightly less than that of the train. Now imagine that the passenger reaches a stanchion, swings around it, and starts walking toward the front of the train. Of course, the velocity of the walker is now slightly greater than that of the train and so, as calculated by an observer sitting on the ground, his or her kinetic energy has increased. There is no magic in this, for the gain in kinetic energy of the walker is balanced by a loss of kinetic energy of the train. Of course one notices no observable change in the train's energy of motion because its mass is vastly greater than that of the stroller. (In this analogy the sitting observer corresponds to the Sun, the train to a planet such as Jupiter, and the passenger to a space probe.)

In any case, by the middle part of the nineteenth century it was generally known to mathematical astronomers that the orbit of a celestial body of insignificant mass could be changed dramatically in certain circumstances. Unfortunately, there were so few engaged in doing celestial mechanics, perhaps a few hundred in the entire world, that such knowledge was inevitably confined to a very small group of scientists. Engineers in general were unaware of these possibilities, and engineers were the ones who for the most part in the 1950s and 1960s were plotting the course of space exploration. At first, relatively few scientists were interested in the space program, and indeed, some were actively hostile.

It was in this context that NASA engineers and scientists, and prominently those at JPL participating in Ranger, Mariner, and Surveyor probes to the Moon, Venus, and Mars, began to examine the phenomenon that came to be known as "gravity assist." Initially, the phenomena associated with gravity assist were regarded as "gravitational perturbations" to Hohmann transfer orbits that affected mission designs and needed to be corrected. But in time, that energy transfer began to be considered as something having a positive and practical application to space exploration.[4]

Interestingly, the idea appeared in the contemporary science fiction literature. Writer Robert A. Heinlein suggested a gravity-assist maneuver as early as 1952 in a novel entitled *The Rolling Stones*. In it a group of adventurers are at a lunar colony and wish to get to a Martian base, but, unfortunately, their rocket vehicle does not have the fuel to make it to that planet by itself. Instead, they take off from the Moon, make a close swing around Earth, and then proceed, flung by Earth's gravity, triumphantly on to Mars. The situation is complicated by the fact that there is a firing of the ship's rocket engine while closest to Earth, but the idea of getting a boost in traveling from one planet to another by means of a close flyby of yet a third world is quite clear.[5]

One distinguished historian of science declared that "the origins of the gravity-assist maneuver are lost in the many and conflicting attempts to determine those origins."[6] The concept of gravitational assist had been known for several centuries, and thus it is not surprising that as the space age approached and finally burst on the scene in the 1950s and 1960s, more and more scientists and engineers began to study its applications to

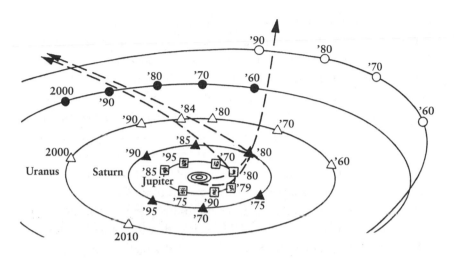

Typical multiplanet trajectories for rare heliocentric geometry of outer planets, 1976–80.
(Long, "To the Outer Planets," 35)

a wide variety of possible missions. One of the possibilities, discussed at JPL and elsewhere, became a part of the developmental and historical processes that led, still later, to the Voyager Grand Tour.

Inasmuch as JPL engineers and scientists were intensely involved in lunar and Mars and Venus planetary probes through the Ranger, Mariner, Mars Voyager, and Surveyor programs, it is not unexpected that they began early to investigate various possible trajectories to those bodies. These initial studies, of necessity, had to consider the three (or more) body problem. Then, in 1961, Michael E. Minovitch, a graduate student in mathematics at the University of California at Los Angeles, came to work at the laboratory as a summer employee. His supervisor, Victor C. Clark Jr., assigned him the task to develop ways to check out the possibilities of various previously calculated trajectories from Earth to Venus and back to Earth again.[7] Minovitch did more. He did not limit his investigations to those trajectories but on his own initiative extended his work to include gravity-assisted transfers between any number of bodies in any order whatsoever. Minovitch is credited as being "the first person to really exploit the concept as applied to non-return trajectories."[8]

Because of the work of Minovitch and others, it became clear that gravitational-assist maneuvers were a practical way to send a spacecraft with a significant payload to a world or worlds that otherwise could not be reached using available rocket launchers, whether these bodies were very close to or very far from the Sun. (It takes a great deal of velocity change, or energy expenditure, to do either from Earth.) Moreover, in many cases such maneuvers can result in drastic reductions in the time needed to carry out such missions. Clearly, NASA was on the way to at least one solution to the problem of time and space involved in interplanetary exploration and travel. Smaller, less powerful rockets than anticipated could efficiently transit the vastness of space to the outer planets and return valuable scientific data to Earth. As a result, NASA began to reconsider its emphasis on building more powerful and larger exploratory space vehicles.[9]

Gravity-assist maneuvers were in fact used first in the early 1970s on the Mariner 10 mission, which used a close flyby of Venus to enable the spacecraft to reach Mercury, and then used the latter planet's gravitational effect to achieve a second and then a third pass by the planet closest to the Sun. Also, at about the same time, Pioneer 11 used Jupiter's gravitational influence to proceed on to Saturn.[10]

Gravitational assist is a general and "flexible" phenomenon. In theory, one can send a space probe first on a direct trajectory to any massive body in the Solar System (comets, asteroids, and most satellites have masses that are too small to have much effect) and then on to another, and another, and yet another, literally without end. There is however, a big drawback to this seemingly rosy picture, for if a probe goes back and forth many times across the Solar System, it is going to take a long time to get to its final destination, and in human affairs, time is of the essence. Mechanical parts wear out over time, power sources become less efficient, miniaturized electronic components are degraded by cosmic rays and other high-energy charged particles, tiny meteoroids blast away at exposed surfaces, and the life-span of the scientists and operating engineers is relatively short. In short, more things go wrong the longer a mission lasts. Thus, even using gravitational assist, it seemed as if several separate, and expensive, missions still would be needed to visit all the outer planets in a reasonably short period of time. Surprisingly, that also turned out not to be the case.

The simple yet elegant solution to reaching all the outer planets in a reasonably short number of years was provided by Gary A. Flandro, then a graduate student in aeronautics at Caltech, who came to work at JPL for the summer of 1965. Flandro's supervisor, Elliott "Joe" Cutting, assigned him "the task of identifying possible unmanned missions to the outer planets." Studying Minovitch's work on the gravity-assist phenomenon convinced Flandro that this "held the key to the outer solar system." Moreover, he realized the crucial role of Jupiter, for its large mass could provide a greater gravitational "boost" than that of any other planet.[11]

Flandro first began to work on trajectories from Earth via Jupiter to Saturn. There had been previous feasibility studies of such orbits, but they were just that—paper studies of feasibility—and, as such, explored few practical details. On the other hand, Flandro had to provide the basis for estimating definite values of such things as "actual flight times, payloads, and planetary approach distances and speeds," as well as identifying the time periods when specific missions could lift off from Earth and accomplish the flights, that is, determining the vital launch periods and firing windows.[12]

In July 1965 Flandro found that the best launch windows for missions to Saturn via Jupiter occurred in the late 1970s, which was just about the time that the necessary booster vehicles, payloads, associated technology, and many other elements needed for successful missions were scheduled to become available. Then he discovered something else. In the early 1980s all of the outer planets would be on the same side of the Sun and quite close to each other as seen from the Sun (or Earth, for that matter). This conjunction meant "that a single spacecraft could explore all four outer planets by using each planet in succession to modify the spacecraft's trajectory as necessary to rendezvous with the next planet in the series." Moreover, as all four giant planets would be on the same side of the Sun, there would be no time-consuming transits back and forth across the far reaches of the Solar System. Thus, missions to all four giants could be carried out in a reasonable time.[13]

What caught technical, and eventually public, interest about this alignment is that it occurs only once every 176 years. The great length of this interval is determined by the motions of the two outermost and slowest moving planets, Uranus and Neptune (excluding Pluto for the moment), for it takes that long for the two of them to again assume their same relative alignment with respect to the Sun. (Saturn, and especially Jupiter, play different roles, for they determine the launch periods.) The last occasion when such an alignment occurred was when Thomas Jefferson was the third president of the United States, and the next would not arrive until the middle of the twenty-second century. This was literally a once-in-a-lifetime opportunity. Of course astronomers of centuries before would have been able to predict this particular alignment, and the idea of proceeding from one planet to the next was not new, but Flandro was in a position to do something definite about those opportunities. And so, he set to work. In a very real sense, Voyager's road to the outer planets was now open, though it would prove to be far from either straight or smooth.

The next year, 1966, Flandro published a technical paper in the journal *Acta Astronautica* that gave detailed descriptions of possible trajectories for a variety of "grand tour" space missions that could visit combinations of Jupiter, Saturn, Uranus, Neptune, and even Pluto. He also included optimum launch dates and sketched the methods by which he achieved the results. He also was careful to emphasize that there was no violation of the law of energy conservation involved. Flandro's study was timely. It became a basic and widely read reference on the subject of multiplanet exploration using the gravity-assist technique.[14]

Interest in the possibility of actually reaching the outer planets with a spacecraft spread widely and rapidly. Perhaps the greatest impetus affecting the dissemination of the idea was a 1966 semipopular (no mathematical equations were included) article by Homer Joe Stewart, who at that time was director of JPL's Advanced Studies Office. Stewart effectively publicized both gravity-assist maneuvers and the golden opportunity offered by the fortuitous positions of all the outer planets in the 1980s. He received publicity and recognition as the inspiration for a NASA project to be called the "Grand Tour" of the outer planets, but as he later wrote, "I have been given a large measure of credit for this work, partly because it was carried out at the Jet Propulsion Laboratory under my general guidance as Manager of Advanced Studies, but, more accurately, because a summary paper which I wrote for *Astronautics and Aeronautics* (December 1966) touched the fancy of the world press and of Congress, and brought about wide-scale discussion on these questions."[15] Thus, the American press and television media in some respects became a catalyst for the conception of a Grand Tour.

Stewart was indeed an effective apostle for Grand Tour missions. His carefully written article was detailed and buttressed by facts, yet it was enthusiastic, easy to read, and totally devoid of pretentiousness. He carefully gave credit to Michael Minovitch and Gary Flandro for their role in the conceptual development of the idea, and to others, as well, in sharp contrast to the often "academic" sounding pronouncements of such August bodies as the Space Science Board (which later became much more effective in communicating with the public). For example, Stewart used the term "interplanetary billiards" instead of "gravity-assist maneuvers" to describe a mission to the outer planets, which analogy, though perhaps not correct in all respects, was close enough and certainly much more understandable to the average American. The U.S. and soon the international press began to give the Grand Tour prominent attention. Subsequently, Congress began to show great interest. Here was a chance to beat the Russians to many planets, not just one, and to do it rather quickly and at no exorbitant cost.[16]

After Stewart's article, the idea of a Grand Tour became something of a media event. Sustaining that basic thrust were the widespread public interest in just what the outer planets and their satellites were like, as well as the thought that people of Earth, and in particular those of the United States, should not miss an opportunity (whatever it might be) that came only every 170 years or so. But a media event can also be misleading.

The Grand Tour idea was intermittently "rediscovered" in the popular media. The August 23, 1968, issue of *Time,* for example, ran a story reporting that the Space Science Board had recommended such a mission and referred to "the new plans." The article then went on to say that "the scientists [of the board] say" that "in 1977 and 1978, planetary positions will enable a spacecraft flying by Jupiter to take a gravity-boosted 'grand tour' that will also take it on past Saturn, Uranus and Neptune."[17] On the basis of this article a reader might reasonably conclude that the Space Science Board originated the Grand Tour concept when, in fact, that was not the case; rather, the board endorsed a concept that was already well and widely known.

Nonetheless, the knowledge that an exploratory scientific mission to the outer planets might be possible did not mean that it was practical. During the 1960s many missions to the outer planets were proposed, but it was clear to those who had knowledge of such things that any such attempts would be much more difficult than probes to such relatively nearby worlds as Venus and Mars. The two main problems still were time and distance.

As for time, even with gravity-assist maneuvers and the fortuitous planetary alignment, years would pass between encounters. It would be like parking a car (and any probe would have to be much more complicated than any automobile) unattended and in the open for years and then betting $1 billion that it would work perfectly when the ignition key finally was turned on—and then losing both the vehicle and the money if it did not. Could individual components, let alone a complicated machine comprising many thousands of them, work for a decade or more? That seemed barely possible in the 1960s. In addition, the efficiency of any conceivable power source would decrease with time due to a variety of effects.

The problem of maintaining a source of power for the exploring machine at such great distances from the Sun was a serious one. Solar cells, which convert sunlight into electricity, had worked well on missions to the Moon, Venus, and Mars, but at Jupiter sunlight is twenty-five times less intense than at Earth, and at Neptune nine hundred times weaker. To generate the needed amount of electrical power, solar-cell arrays used to gather that power would have to be impossibly large and massive. In addition, the weaker sunlight also meant that any imaging devices having to deal with a planet or satellites whose surfaces appeared some nine hundred times darker at Neptune than at Earth, required correspondingly longer exposures, necessitating additional expenditures of power simply to compensate for the relative motion of the probe and celestial body during those longer intervals.

Another problem associated with greater distances from the Sun, and the correspondingly weaker intensity of sunlight, is that of colder temperatures. Unfortunately, most things on board a scientific probe, including mechanical parts, lubricants, computers, electronic components, propellants, and various sensors, have to be kept within reasonable temperature limits, requiring additional expenditures of precious energy.

Greater distances also meant that radio signals either from Earth to a probe, or from a probe back to Earth, are weaker when received. To command the spacecraft and also receive data from it at great distances would require some combination of more powerful transmitters, more sensitive receivers, and larger antennas than needed on missions to the inner Solar System. Moreover, it takes a signal traveling at the speed of light four hours to reach a spacecraft at Neptune, and four more hours for a response to arrive back on Earth. Thus, it is impossible to control a spacecraft at that distance in "real time"; a tour of the outer planets required minute planning and forecasting. The technical problems of a Grand Tour were enormously more complex than those involved in exploring the Moon and nearby planets. Navigating between the planets presented a daunting challenge; to ensure that the encounters occurred at the right places and the right times over a voyage of many years was no easy task.

Still other problems abounded. For example, would passage through the asteroid belt between Mars and Jupiter disable or even destroy a probe? And what damage might be done during a passage through Jupiter's intense radiation belts (and possibly those of the other giant planets)? Similarly, might passage through a planet's ring system demolish a spacecraft?

Finally, there were myriad detailed engineering problems: how to design Sun sensors to operate under greatly varying solar illumination, how to control on-board thrusters to provide exactly the change in velocity that is needed (and how to calculate and measure that velocity change), how to design and program the on-board computers that would control the spacecraft during the short periods during planetary encounters when Earth commands cannot reach the vehicle in time to make needed changes, how to recover if a meteoroid strike sets the spacecraft spinning, and what to do if a major earthquake or other disaster on Earth disrupts communications. And what if Congress fails to provide funds to continue operations a decade or so in the future? The "what if" list was lengthy, but much more than idle speculation.

Whatever missions eventually would be flown, problems had to be faced well in advance. Aerospace engineers, and in particular those at JPL, knew this well by past experiences with Ranger, Surveyor, Mariner, Mars Voyager, and other programs. Thus, JPL began to conduct a large number of design studies to define just what kind of spacecraft could carry out missions to the outer planets, and whether or not the necessary technology (launch vehicles, computers, scientific instruments, tracking facilities, and the like) was available or could be developed in time to be used on a Grand Tour mission.

In large part these were "paper studies" that did not include the building of actual prototype mission "hardware." They were meant to define, as accurately and precisely as possible, what could or could not be done, and what were the best choices among available alternatives for performing successful missions. Among the related Grand Tour studies was the Double Precision Orbit Determination Program (DPODP). John R. Strand, one of the forty or so contract programmers working on that program and interplanetary trajectory problems at JPL between 1966 and 1970, explained that as compared to a single planet encounter, "targeting would be intensified by orders of magnitude with a multi-planet slingshot energy transfer."[18] Thus, although the original Grand Tour program never came to fruition, associated trajectory work such as DPODP contributed to a great leap in navigation accuracy that, in turn, made Voyager missions possible—and ultimately successful.

Most of this work received little public attention but is summarized in a progress report published in 1969 by James E. Long of JPL's Office of Plans and Programs. Long personally directed a study during 1967–68 of a proposed 1972 Jupiter flyby mission. In his report, Long offers a succinct yet comprehensive account of the results of JPL "grand tour—outer planets" studies conducted up to that time.[19] Those studies became a part of the history and the technology of what would one day become the Voyager Grand Tour of the outer planets.

One of the problems with gravity-assisted missions is that there is a wide variety of destinations from which to choose. Long described specific options that had been investigated in some detail: (1) a flight to Jupiter, Saturn, Uranus, and Neptune, with possible launch dates from 1976 to 1979, 1977 and 1978 providing the best opportunities; (2) a flight to Jupiter, Saturn, and Pluto, again viable from 1976 to 1979, with optimal launches in 1977 and 1978; and (3) a flight to Jupiter, Uranus, and Neptune, with launch windows extending from 1977 to 1980.[20]

A number of factors affect the specific trajectory chosen to go from one planet to another. First, any such path must ensure that the probe will get to the next planet in the sequence of encounters. This, in turn, means that at each encounter the spacecraft must have some definite yet achievable combination of approach velocity (speed and direction of motion at a specified distance from the planet) and location and distance of closest approach. In addition, the scientific requirements of the mission (which, after all, are the main reasons for going in the first place) demand certain "miss" distances, viewing angles, and so forth. Finally, any close pass by Saturn must avoid its rings (later, the discovery of rings around Uranus and Neptune would similarly complicate flight trajectories in their vicinity).

As a practical matter, it appeared necessary to make small adjustments to a probe's velocity (by means of an on-board rocket engine) both before and after an encounter. To do this with the necessary precision required not only radio tracking data but also something like an on-board imaging system to determine the apparent position of the planet with respect to the background stars.

A most important consideration in any exploration of the outer planets was the type of on-board scientific instruments to be included. The nature, size, mass, and energy consumption required fundamentally affected the design of the spacecraft and the design of the mission. Because these early studies did not consider actual, approved missions, there was a wide choice of instruments available. Several groups of instruments were considered, but in the main, they comprised what had flown on earlier planetary probes. That is, they included devices to take images, study planets in infrared and ultraviolet wavelengths (which are not observable from Earth's surface), measure magnetic fields and the abundance and energies of charged particles such as electrons and protons, and so on.

JPL mission planners focused on the detailed design of a Grand Tour mission to Jupiter, Saturn, Uranus, and Neptune. During such a trip a probe would be, most of the time, in an interplanetary "cruise mode" with not much happening. During those intervals the spacecraft would be "listened to" by its operators only on occasions—perhaps every two weeks or so. That strategy would save a lot of operating time and money, but it meant also that the probe would have to have a number of on-board capabilities, such as a data-storage system and a way to control its operations and carry them out at specific times, plus have the ability to be "reprogrammed" and redirected from Earth if necessary.[21]

As to the question of how best to supply the electrical power that a mission to the outer planets required, JPL studies considered a variety of possibilities but came down decidedly in favor of using radioisotope thermoelectric generators (RTGs). Such devices converted the energy of particles given off by radioactive isotopes into heat, which was then converted into electricity. They had the advantage of being relatively small, lighter than solar arrays, and independent of distance from the Sun. On the other hand, they had disadvantages, such as the effects their radiation might have on spacecraft components (especially on electronics and instruments designed to study charged particles), not to mention the international complications that might occur if the launch vehicle failed and radioactive debris was strewn over a wide area. To be sure, RTGs were designed to survive launch accidents.[22]

Consideration was also given to hazards such as those due to passage through the asteroid belt (which turned out to be negligible) and Jupiter's radiation belt (these turned out to be severe). Reliability, of course, was a prime concern. There were several ways to help ensure longevity. One was to maintain tight quality control over the design, fabrication, and inspection of all components, another was to provide redundancy (a backup for critical systems), and a third was to make it possible to "work around" less serious failures, even in flight.

When it came to the actual physical shape of the spacecraft, and the placement of its individual systems, designers faced some hard choices, many of which involved trade-offs between various sets of ideal circumstances. For example, a "high-gain" antenna for communications and control had to point toward Earth most of the time, which affected temperatures in various locations of the spacecraft. As another example affecting design, an instrument that measured magnetic fields had to be physically far from any magnetic substances on the spacecraft itself, and the radioactive RTGs had to be as far as possible from everything. Meanwhile, instruments that were to investigate a planet or satellite (such as an imaging system) must be placed so that, for example, the spacecraft itself would not block their view. Add to these considerations factors such as, for example, a substantial mass on one side of the vehicle must be balanced by another on the other side to avoid undue stresses during launch and "midcourse maneuvers" and it becomes obvious that a very delicate "juggling act" is involved in the design of a scientific probe. There was no perfect solution. Designers were faced with choosing the best of several possible alternatives.

Naturally, all this effort would have been for naught if no booster rocket with the necessary performance were available to send a probe to Jupiter. JPL engineers determined that a spacecraft weight of slightly over one thousand pounds would be needed, requiring a very large launch vehicle. Saturn rockets with modified or additional upper stages met the requirement, but they were very expensive. Moreover, as the lunar program progressed and congressional expenditures on space tightened, it had become clear that Saturn V rocket production would soon be halted.

Fortunately, the Air Force's Titan III booster was available and held promise of being available for some time to come, as it was scheduled to launch various reconnaissance payloads vital to national security. A modified version of this rocket, incorporating a Centaur as an upper stage, along with an additional solid-fuel rocket (the Burner 2), could do the job. Eventually, such a rocket assembly launched the Voyagers.

When it came to "putting it all together," JPL engineers had to look at a variety of compromises. There were four preferred spacecraft designs, but one seemed superior. This was a vehicle that was stabilized in three axes with respect to the stars, its orientation maintained by sensors that looked at the Sun and Canopus, a bright star far from the main plane of the Solar System. Gyroscopes controlled the orientation during the intervals when changes in attitude had to be made. Radioisotope thermoelectric generators powered by the natural radioactive decay of plutonium were mounted on a boom to provide the power, and a high-gain antenna was folded up during launch and unfurled afterward. Science instruments that needed to be pointed in various directions during a flyby were mounted on a gimbaled platform, and sensitive devices with special requirements, such as those that measured magnetic fields, were mounted on other booms.[23]

There were subtle but important "interface" problems as well. For example, would some commands to the spacecraft from operators on Earth interfere with or contradict those commands generated on board the spacecraft? Could the available electrical power be distributed effectively to where it was needed at all times? Would various maneuvers be com-

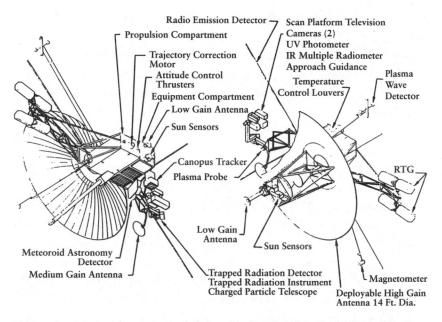

Radio Emission Detector
Propulsion Compartment
Trajectory Correction Motor
Attitude Control Thrusters
Equipment Compartment
Low Gain Antenna
Sun Sensors
Scan Platform Television Cameras (2)
UV Photometer
IR Multiple Radiometer
Approach Guidance
Temperature Control Louvers
Plasma Wave Detector
Canopus Tracker
Plasma Probe
RTG
Low Gain Antenna
Sun Sensors
Meteoroid Astronomy Detector
Medium Gain Antenna
Trapped Radiation Detector
Trapped Radiation Instrument
Charged Particle Telescope
Magnetometer
Deployable High Gain Antenna 14 Ft. Dia.

Thermoelectric outer planet spacecraft designed by TOPS. (*Space Daily,* November 19, 1970, 79)

patible with one another? Would electrical currents generated somewhere on the spacecraft produce effects that might interfere with other systems? These were only preliminary studies but, with hindsight, one can clearly see the basic outlines of what became the Voyager space probes emerging from the Grand Tour studies conducted at JPL primarily between 1966 and 1968.

Useful and informative as the early Grand Tour studies were, by the late 1960s it had become clear that more detailed, and more practical, design and development work was in order. The result was TOPS, the Thermoelectric Outer Planet Spacecraft program, which began at JPL in 1968. TOPS was never a flight program, but instead, an advanced systems technology, or engineering technology, project. It grew out of the earlier Grand Tour studies. Over the years there has been some confusion because TOPS could be taken to stand (mistakenly) for "Tour of the Outer Planetary System" (which it definitely was not) and because many elements of the TOPS design appeared later in the Voyager probes. Indeed, the Voyagers as flown had a strong resemblance to the TOPS "baseline" spacecraft.

The TOPS program differed from earlier JPL work in that it was a single, unified project, not a series of separate studies, and in that it included actual hardware design and fabrication, or "breadboarding," of selected components. This did not mean building equipment meant for flight but the creation of versions of spacecraft subsystems, and even some systems, that could be tested to see if and how they worked. Although there never was an actual TOPS mission, or even a physical TOPS vehicle, it nevertheless was a vital predecessor to Voyager. As with the earlier Grand Tour studies, most of the TOPS results appeared in various JPL publications that were mainly intended for internal use, or for prospective contractors.[24]

In general, the TOPS project proceeded in a fashion similar to those of the earlier studies described above, for no definite mission or missions to the outer planets had been authorized. Thus, TOPS planners considered a variety of possible solutions to a myriad of problems. In addition, TOPS engineers could apply detailed and practical information such as test results on actual physical components that might be used on a spacecraft targeted to the outer planets.

As a result, by 1970 TOPS planning and the design of missions to the outer planets had taken form but lacked content. A NASA-approved project, called Grand Tour, provided the content. In effect, TOPS constituted an essential planning regimen for a proposed Grand Tour of the outer planets. NASA assigned management responsibility for a program entitled Outer Planets Grand Tours to JPL, and the laboratory established a Grand Tour Project Office and appointed a preliminary mission design team. In October 1970 NASA issued a call for proposals for the development of science experiments for the proposed program, and in November engineers and managers conducted a preproposal briefing for potential contractors. It was to be a mission that fired the public imagination and tested NASA's engineering ingenuity. But, as Bruce C. Murray, a Caltech professor of planetary science who later became director of the Jet Propulsion Laboratory, observed, "a funny thing happened on the way to Neptune"—and Jupiter, Saturn, and Uranus.[25]

3

A FUNNY THING HAPPENED ON THE WAY TO NEPTUNE

As had been the case for the Voyager Mars program, the dynamics behind JPL's proposal for a Grand Tour of the outer planets had much to do with the deep conviction among scientists and engineers at the laboratory that exploration of the Solar System using scientific probes offered the most promising opportunity for America in space. "Our ignorance of the more distant portions of this solar system is so great," Bruce Murray wrote, "the potential scientific discoveries so varied and unpredictable, and the potential cultural impact so great and enduring, that outer planet exploration is, perhaps, still the best new investment for the U.S. space program available." When a program was approved for exploration beyond Mars, Murray urged, "it should be prosecuted vigorously and imaginatively."[1]

In February 1968, Murray testified forcefully on behalf of a Grand Tour of the outer planets before the House Subcommittee on Space Science and Applications. The Soviets, he pointed out, had been the first to successfully put an exploring vehicle into orbit, and recently, in October 1967, the Soviets had been the first to land an exploratory spacecraft on the Moon. American scientists had come to realize that they might learn more about the surface conditions and atmospheric composition of Mars through Soviet news reports than as a result of their own efforts. Murray anticipated the possibility of an advanced type of Soviet mission to Mars as early as 1969. However, he suggested that U.S. superiority in communications and spacecraft reliability justified American efforts to explore Mars and conduct flyby missions to Mercury and Jupiter. These missions, coupled with preparations to exploit the "once-in-a-century" opportunity in 1976, 1977, or 1978 to take the Grand Tour of Jupiter, Saturn, Uranus, and Neptune, could be accomplished and would reestablish American leadership in the scientific exploration of space with no commitment to major funding increases.[2]

Harris M. "Bud" Schurmeier, a key manager in the development of the Ranger and Mariner spacecraft, became JPL's deputy assistant director for Flight Projects in 1969 and, like Bruce Murray, persistently voiced the very deep commitment of JPL engineers and scientists to planetary exploration. Schurmeier suggested that the dynamics of space exploration proceeded from four imperatives: (1) scientific interests and objectives, (2) technical capability, (3) the natural environment, and (4) economics. There had to be, he elaborated, a scientific motive or need plus the technical-engineering ability to send a vehicle into space, and also there had to be adequate financial resources. Science objectives and economics interacted to frame the mission. The technical capability of the spacecraft and the physical properties and uncertainties of the universe defined and determined the ultimate success of the mission. Only in the best of times, he said, could pure science compete successfully for national attention in the economic sense. Thus, "planetary exploration has inevitably been insignificant on a national scale, low on the agenda, small in the budget."[3] But Schurmeier envisioned the Grand Tour as the space "mission-opportunity" for the next decade and the "After Apollo" epilogue for the lunar landing program.

The idea of a Grand Tour of the outer planets had been planted and was "vigorously and imaginatively" nurtured in the rich and singular seed bed of JPL/Caltech technology and culture. Robert Parks, Bill Pickering, Mike Minovitch, Gary Flandro, Homer Joe Stewart, James E. Long, Bud Schurmeier, Bruce Murray, and, effectively, the entire JPL/Caltech society breathed life into the Grand Tour. In March 1969, Donald P. Hearth,

NASA's director of planetary programs in the Office of Space Science, informed the House Subcommittee on Space Science and Applications, before which Murray had appeared the previous year, that NASA had assigned "high priority" to a future exploratory mission to the outer planets that was made possible by the unique alignment of Jupiter, Saturn, Uranus, Neptune, and Pluto in the 1976–80 time frame. In June 1969 *Astronautics & Aeronautics* featured an article by JPL's James E. Long entitled "To the Outer Planets," which provided an informed overview of the conceptual development of the proposed mission and described its technical characteristics.[4]

Fortuitously for the cause of planetary exploration, JPL's Mariner 6 on July 31, 1969, followed by Mariner 7 on August 5, entered Martian space and began returning scientific data and over two hundred televised photographs of Mar's surface from a minimum altitude of only a few thousand miles. It had taken Mariner 6 156 days and Mariner 7 130 to complete the 241-million-mile flight. The twin Mars-Mariner flights were the first successful dual-flight scientific explorers. Launched from Cape Kennedy, Florida, by Atlas-Centaur launch vehicles on February 24 and March 27, they were tracked and directed to their targets by the NASA-JPL Deep Space Network. On the occasion of their launch, the *Wall Street Journal* had highlighted the venture as an effort to solve the great mystery of interplanetary life: "Are there living things on other planets? Or is the earth unique in the solar system, and perhaps in the universe, as a cradle of life?" Those were the big questions, staff reporter William E. Blundell explained, and the solutions lay not just in the exploration of Mars, but in the extension of that exploration by a "grand tour of four of the five planets that are farthest from the Sun."[5]

Interestingly, Blundell offered the "complexity and precision" of the twin Mars-Mariner flights as a "curious contrast" to the operations of the Jet Propulsion Laboratory: "JPL is a place where the bosses are forever trying to get the workers to slow down a little, where the organization chart looks like a management consultant's nightmare and where absent-minded scientists have been known to wander around in their pajamas in the middle of the night."[6] That news article, and many others, plus science journals and discussions relating to the Grand Tour, served to pique both public and professional interest in JPL and in the proposed tour. Even before the twin Mars explorers had arrived at their destination, Mariners 6 and 7 were announced by the media as being precursors to a Grand Tour of the outer planets. Their eminently successful missions reinforced a growing political and public consensus that there should indeed be a Grand Tour, and that JPL could accomplish the mission if it were attempted.

The Space Science and Technology Panel of the President's Science Advisory Committee, which included Murray in its membership, offered a belated and a guarded endorsement of an exploratory mission to the outer planets in March 1970. The panel recommended an expanded program of Earth-oriented research and related applications of space science and technology and the exploration of the Solar System with emphasis on the Moon, the search for extraterrestrial life, and a "diversified program of planetary exploration." For the scientists, the basic question was understanding the evolution of the Solar System. The panel suggested that a key to that understanding involved the "massive outer planets." Interesting facets of outer-planet exploration included a study of the planets' "varied and complex satellite systems" and their chemical and physical properties. The magnetic fields of Jupiter and Saturn needed attention, as did Saturn's rings. Little was known about Uranus and Neptune, and even less about Pluto, but the panel seemed to attach little urgency to their exploration.[7] The emphasis on outer planet exploration by the Space Science and Technology Panel, such as it was, stopped at Jupiter and Saturn.

President Richard M. Nixon was much less equivocal than his science panel in endorsing a Grand Tour of the outer planets. "During the next decade," he announced on March 7, 1970, "we will also launch unmanned spacecraft to all the planets of our solar system. . . . In the late 1970s, the 'Grand Tour' missions will study the mysterious outer planets of the solar system—Jupiter, Saturn, Uranus, Neptune, and Pluto. The positions of the planets at that time will give us a unique opportunity to launch missions which can visit several of them on a single flight of over three billion miles. Preparations for this program will begin in 1972."[8] With the president's unequivocal endorsement behind them, in April 1971, JPL released its new five-year plan, which featured an outer planets exploration program.

The overall guidelines for program development included the primary objective of continuing to lead the scientific exploration of the Solar System, and continuing to advance the technologies needed for scientific exploration in space. As of 1971, programs ongoing at JPL included a Mariner Mars 1971 flight, a Mariner Venus/Mercury flight for 1973, the Viking Mars lander program scheduled for launch in 1975 in cooperation with Langley Research Center, and tracking and data acquisition for most of the existing NASA manned and unmanned space flights. JPL's major anticipated new program effort was the Grand Tour.[9]

There were, however, some cautionary notes. The Space Science Board, in its 1970 summer study, advised against "large, focused, expensive projects" such as the Mars Voyager program (recently eliminated) and the Grand Tour missions. Although these missions might have good scientific returns, they did not fit within a realistic assessment of the NASA budget situation according to the Space Science Board, and if initiated would result in curtailing many of the (presumably more productive) "small-scale, low-cost" science activities. A "JPL Discreet" program memorandum signed by Pickering thus advised that efforts were underway to limit outer planet exploration to Jupiter alone, on the assumption it would be less costly. Pickering questioned the validity of that assumption.[10]

At the core of the brewing controversy among scientists over prospective funding for the Grand Tour was the fundamental issue that such funding would presumably deplete NASA financing for other (non-JPL) projects. Headquarter's most conservative preliminary Grand Tour spacecraft cost estimates, for example, ranged from $750 to $900 million, and related science experiment budget requests exceeded $2.5 million. In addition, as NASA science programs began to take form, NASA research centers usually began to compete for the management or "lead center" role for those programs. Space science could be keenly competitive—sometimes, Bruce Murray thought, ruinously so.[11]

An experienced team headed by Pickering and Bob Parks pitched a full presentation to NASA's Office of Space Sciences and Applications in June 1970 to obtain a "project start" authorization for the Grand Tour and initial funding for 1972. Broad mission objectives were to investigate the environment, atmospheric properties, surface characteristics, and body properties for each of the five outer most planets and their satellites. Grand Tour science would focus on illuminating the known unusual features of each planet, and the study of the properties of the solar wind plasma and energetic particle environment throughout the outer Solar System. More specifically, Donald G. Rea, deputy director of planetary programs, advised that the missions were to study Jupiter and its moon Io; to examine Jupiter's radiation belts, polar regions, and mysterious "Great Red Spot"; and to investigate Saturn's rings, Titan's atmosphere, Uranus' inclined axis, and the character of true interstellar space.[12]

The underlying scientific rationale for the Grand Tour was the "meagerness" of our knowledge of the outer planets. Thus, Grand Tour was to be truly an exploratory mission, "one in which the most important results might be ones not possible to predict or even imagine with our present understandings."[13] The real science justification was to search for the unknown—a position that immediately raised the question of costs, payoff, mission options, and program alternatives.

Ronald F. Draper, for example, who was a member of the Grand Tour presentation team and had been the spacecraft system engineer on the TOPS program, reflected later that one of the reality checks of the proposed Grand Tour had to do simply with the fact that previously no spacecraft had been built for a flight lasting three and a half or more years. There was a very real question about reliability. There were few, if any, machines on Earth that could operate for years without maintenance and repair, much less machines that could function for extended times in the largely unknown environment of space. Moreover, even the most advanced RTG energy packages of the time lost substantial power after ten thousand hours. The Thermoelectric Outer Planet Spacecraft design project just completed provided the critical technology that made the Grand Tour possible, Draper said, and was a direct precursor.[14]

After working on spacecraft engineering design for Mariner 3, 4, and 5, Draper moved to the TOPS project in June 1967, working as the spacecraft system engineer and manager of the design team under the project manager, William S. Shipley. The project made significant contributions to the development of new, advanced spacecraft technology. It was, for example, the first spacecraft design to use the RTG. The design concept was turned over to the Department of Energy for development, which in turn cooperated with industry. Indeed, because of TOPS, industry became interested in the development of various advanced concepts for spacecraft flight and design. Ford Aerospace, Philco Electronics, Watkins-Johnson, Hughes Aircraft, and Motorola, for example, all made contributions to the development of the communication systems, including antennas and S- and X-band traveling wave tubes (TWTs). Then, too, insulation materials were developed that extended the lifetime of the thermoelectric power sources.[15]

One of the most serious problems addressed during TOPS development was related to the proposed long-term Grand Tour operations and concerned the computer control subsystem (CCS). Computers had to be able to navigate the spacecraft independently of control from Earth. In addition, the spacecraft computers not only needed to be able to detect subsystem failures and initiate corrective actions but also had to be capable of accepting reprogramming and directions from Earth. In other words, the Grand Tour computer systems needed to be "self-test-and-repair" (STAR) general-purpose machines and use a minimal amount of power. Luckily, STAR systems were, in fact, under development independently of both the TOPS and Grand Tour programs. As Draper explained it, the STAR system essentially involved the use of five independent computer control units, each "voting" on a decision. A common decision by any three of the control units overrode the action of a presumably defective one. Ultimately, STAR influenced Grand Tour design, but the computer system was not incorporated into the later Voyager Grand Tour design because of costs.[16]

Given the available technical design capability, what was to be the actual overall Grand Tour mission design? Draper, Roger D. Bourke, and Charles F. Mohl described a "baseline" mission including a Jupiter orbital explorer to be launched in 1975, two Jupiter-Saturn-Pluto missions scheduled for 1977, and two Jupiter-Uranus-Neptune probes for 1979. An "alternate mission set" could include a single launch in each year, 1975 to 1979. Mission planning included considerations relating to the closest approach of target planets to Earth, optimizing encounter periods, and potential schedule conflicts with Viking and other flight programs.[17]

Bill Shipley addressed problems relating to longevity, instrument reliability, and radiation damage, using Mariner and TOPS experiences as benchmark indicators. He also discussed ramifications of building a spacecraft and scientific instruments for a twelve-year operating life when there was no way components could be subjected to a twelve-year time test. Charles W. Cole reviewed scheduling, gave a management profile, and described existing resources and procurement policies.[18] Pending congressional budget action, NASA approved Grand Tour.

JPL then established a Grand Tour Project Office and appointed Bud Schurmeier project manager. In turn, Schurmeier put Draper in charge of the Engineering Design Team and asked Raymond Leroy Heacock, who had worked with him on Mariners 6 and 7, to be the Spacecraft Systems Manager for the project. Heacock had earned undergraduate and master's degrees in electrical engineering at Caltech before going to work at JPL in June 1953. As had so many others at JPL, he worked his way up and through the Army's Corporal, Sergeant, and Pioneer programs, and then, with the transfer of JPL to NASA, worked on instrumentation and imaging for the Ranger and Mariner programs. Then he was a project representative on Mariner 6 and 7 before moving to Grand Tour.[19] In such ways, Grand Tour technology and Grand Tour management derived from JPL's historical experience.

By spring 1971, more than 500 scientists had offered competitive proposals for experiments to include on the Grand Tour missions. From those, by early April, NASA and JPL selected 108 scientists representing thirty-six different educational institutions and laboratories from the United States and scientists from six foreign countries (Canada, Denmark, Germany, France, Sweden, and the United Kingdom) to participate in the definition phase of the outer planets missions. Those scientists were grouped into thirteen teams, with responsibility for imaging, radio science, plasma wave, photopolarimetry, energetic particles, plasma, ultraviolet spectroscopy, planetary radio science, astronomy, infrared spectroscopy, magnetic fields, meteoroids, atomic hydrogen measurements, and planetary x-rays. Each team had a group leader who also served on the Outer Planets Mission Steering Group responsible for integrating the individual science experiments into the overall missions.[20]

Grand Tour science, by every standard, was to reflect the best and most advanced science and technology; but that science and technology were also costly. The tour's estimated costs, originally projected to be in the range of $500–615 million, were now almost doubled. In April 1971, the Office of Management and Budget (OMB) asked NASA's Office of Space

Science and Applications to devise a simpler, less costly alternative to the proposed outer planets mission. Associate Administrator John Naugle assigned Schurmeier, Warren Keller, and Edward Wash to work with OMB representatives on the case. Naugle thought that NASA should prepare comparative cost options, including (1) a modified Pioneer spacecraft, (2) a distinctive outer planets spacecraft, (3) the TOPS design, and (4) a "modified" TOPS. Comparative cost estimates would provide a basis for critical analysis of the Grand Tour project. Cost reductions, Naugle stressed, had to be related to losses in science capability and increases in mission risk.[21]

In May, NASA and JPL investigated the possibility of substituting the proposed NERVA (nuclear engine for rocket applications) nuclear launch vehicle, then in the design stage, for a Titan IIID–Centaur launch being planned. They also explored the possibility of using a modified "fat" Centaur launch vehicle. With NERVA, the trip time to Pluto and Neptune could be reduced from the projected 8.5 to 9 years to 6 years or less, and bypassing Jupiter could reduce the time in flight even more. Shortening the operations time significantly reduced costs, but the cost of a "fat" or modified Centaur could not be established until the actual vehicle configuration and mission and fuel requirements were set.[22] "Costing" a space mission, particularly one to the outer planets that could extend for over a decade, was at best problematical.

By May 1971, budget requests for science projects related to the Grand Tour had exceeded $2.5 million, while the project budget recommendations for science approached $1.5 million. Growing restlessness, if not outright opposition, to Grand Tour costs developed in Congress and within the science community. The House of Representatives approved $30 million for "project-start" work on the Grand Tour for fiscal 1972, but the Senate reduced that authorization to $10 million. The Senate, cognizant of the rising costs of the war in Vietnam and domestic programs, plus spiraling inflation, suggested that NASA reconsider the Grand Tour program, opting perhaps for a two-planet mission and possibly the "later" use of a NERVA nuclear rocket for outer planet exploration missions.[23]

The Jet Propulsion Laboratory, optimistically, hopefully, and either with great insensitivity or its usual intractable determination, defended a full-scale, TOPS-type outer planets Grand Tour mission. Congress, for a moment, seemed to relent in its rising opposition to Grand Tour by approving a compromise $20 million for fiscal 1972 but then going on record with the promise that NASA would eventually get the full $30 million. But unmistakably, there was a growing sense that the Grand Tour program was too ambitious and its costs too high—by July 1971, estimates ran from $856 million to $1 billion.[24]

Costs compounded other complications. Support for the Grand Tour concept among many scientists had from the first been tepid. Now, with cost estimates rising, that fragile approbation had been transformed into positive opposition. Clearly, Grand Tour would deflect funding from other, possibly more worthwhile scientific studies. In addition, some members of the scientific community had, under the duress of withering funding from NASA for scientific projects, become actively opposed to continued or expanded funding for manned space flight; they reasoned that as manned space flight consumed the greater part of the NASA budget, it left little for science. James A. Van Allen, for example, whose radiation experiment had first flown aboard Explorer 1 in January 1958, and Thomas Gold, a member of the American Geophysical Union, vigorously and actively opposed any extension or continuance of manned space programs—including the development of a proposed space shuttle. They advocated a substantive reduction in the space budget, with most of the resources devoted to applications and science projects.[25]

That belligerent opposition to manned space flight programs by members of the science community, albeit a minority of that group, generated a "counter-insurgency" effort against robotic science programs. Thus, NASA's *Space Business Daily* observed in July 1971 that serious opposition to the Grand Tour stemmed in part from "the repetitious congressional appearances of anti-manned space flight leaders and the weighty exposure given their arguments within the media, such as the recent appearances of Van Allen and Gold, has had the effect of losing support for the unmanned space exploration movement."[26] The arguments and opposition by scientists to human space flight in general, and to expensive outer planets exploration programs such as Grand Tour, were seized upon by very vocal elements in American society who either believed that the United States had no business in space whatsoever, or that funds allocated to space would be better spent at home on Earth for medical programs, environmental programs, education, highways, defense, music, art, or other programs that would improve the general welfare; opposition to space spending of any kind rose markedly.

Although the Space Science Board of the National Academy of Sciences submitted a new report in November 1971 that described the extensive study of the outer Solar System as "one of the major objectives of space science in this decade," it recommended funding for the program at less than half of the current projected levels, or about $400 million for the decade. That meant that TOPS-type spacecraft launches would be reduced to two rather than four or five, and that the science payloads would be limited to 130 pounds for each flight; or that NASA must redirect its outer planets effort to use Pioneer or Mariner-type spacecraft. In addition, the Space Science Board urged a strong program of Earth-based studies of the outer planets using satellites, rockets, balloons, aircraft, and ground-based instruments. Some scientists thought that a large, 45-inch Earth-orbiting telescope would produce more good science than a tremendously more expensive Grand Tour.[27]

Conversely, the Physical Sciences Committee of NASA's Space Program Advisory Council unanimously recommended a TOPS-type Grand Tour in four launches, but it also suggested that because of the greater simplicity and lower costs associated with exploring the more accessible Jupiter and Saturn, an independent program be developed by NASA for that purpose. In addition, the council supported the continuing development of improved launch vehicles such as the Titan with a seven-segment solid booster. The committee also cautioned against the weight and costs associated with imaging and television in an outer planets exploratory craft and considered a study of the chemical composition of the outer planets to be the major contribution that could be made by a Grand Tour mission.[28] Thus, even while the mainstream science establishment seemed to support continuance of the Grand Tour, the signals were uncertain.

In addition, there were ramifications beyond the science community. Grand Tour, Viking, space shuttle development, modification of the Centaur, the NERVA nuclear rocket program, the human space program (including the ongoing Apollo lunar program), and robotic exploration of the Solar System were all interrelated in terms of technology and in terms of costs. Wernher von Braun, for example, at a meeting of NASA associate administrators in September 1971, worried that the NERVA program competed for funds needed for outer planets missions. Congress, he suggested, seemed to have a special interest in NERVA, even though NERVA had not been associated with any real program or objective. He feared the loss of the Grand Tour to a NERVA that had no place to go.[29]

Similarly, engineering studies for the development of a seven-segment solid booster for the Titan III threatened to disrupt the Viking program. That exploratory effort, which came into being after the cancellation of the Mars Voyager project, proposed to orbit and soft-land spacecraft loaded with scientific packages and television cameras on Mars. Viking was scheduled for launch by Titan-Centaur vehicles, but in 1969, money problems caused a postponement of such launches from 1973 to 1975. NASA remained concerned that any additional delays or diversions could force the closure of the Viking program.[30]

Following the successful lunar landings in 1971 by Apollo 14 with astronauts Alan B. Shepherd, Stuart A. Roosa, and Edgar D. Mitchell aboard, and then by Apollo 15 with David R. Scott, Alfred M. Worden, and James B. Irwin (which included the first EVA using a lunar roving vehicle), NASA scheduled Apollo 16 and 17 for the following year. But the Office of Management and Budget felt the need to reduce NASA's budget by $500 million and singled out the elimination of the Apollo 17 lunar flight as an off-the-top $100 million savings. The same budget actions that would delete Apollo 17, however, could also force a severe setback in Grand Tour mission planning and development. In part, the question was, should NASA fight for retention of Apollo 17, culminating its tremendously successful lunar landing program, and also work for the continuance of Grand Tour, possibly losing both? Or should NASA defend one and not the other? In October 1971, NASA administrators estimated the probability of receiving funding for the Apollo 17 flight at fifty-fifty.[31] Grand Tour funding was not scheduled to begin until 1972.

To be sure, there was much more at stake than the last flight of Apollo or the development phase of the Grand Tour program. By 1970, following some years of intensive study, NASA had determined that the future of America in space, following the completion of the Apollo lunar flights, should be the development of a reusable orbital spacecraft. Such a spacecraft was viewed, in part, as a cost-savings program. Preliminary design work in 1969 and 1970 led to the decision reached by June 1971, to substitute for a fully reusable orbiter—one that would be placed in orbit by expendable booster rockets and tanks but would return to Earth under its own power. The costs for developing the "partially" reusable vehicle were much less than the costs of a fully reusable spacecraft.[32]

As of 1971, however, Congress had provided no specific funding for space shuttle development nor, indeed, had Congress provided funding for the Grand Tour. Meanwhile, NASA shuttle program expenditures had leaped from $12.5 million in 1970 to $78.5 million in 1971. Thus, at the beginning of 1971, NASA was at a critical juncture in obtaining funding for the completion of the Apollo program, for continuing Viking, for developing the Grand Tour, and for the inception of the space shuttle.[33] As never before, costs drove the American space program and choices had to be made. Those choices, however, were not JPL's to make, nor wholly those of NASA.

Grand Tour, NASA, and JPL contractors continued to plan for a Grand Tour of the outer planets. In December 1971, North American Rockwell engineers conducted a three-day, JPL-sponsored study of possible particle radiation effects on the Grand Tour. About 150 scientists and engineers from various NASA centers, and from universities and laboratories involved in the development of Grand Tour science experiments, attended. The emphasis was on how to guard science packages and the spacecraft from the adverse effects of such radiation.[34]

In the week following the close of the radiation conference there was good news and bad news for both NASA and the Grand Tour. In January 1972 President Nixon announced his support for the space shuttle. That decision, followed by congressional budget support for the shuttle, taken in the context of escalating costs of the war and double-digit inflation, meant that NASA's diminishing funds would continue to be heavily committed to manned space flight. The Office of Management and Budget accepted NASA's final request for $3.2 billion for fiscal year 1973, after having countered the original request for $3.6 billion with a $2.8 billion "guideline." This compromise budget figure would provide funding to complete the final two Apollo flights, but NASA's $50 million funding request for the Grand Tour was going to be cut to $30 million under the budget compromise. "Some say," *Space Business Daily* noted, "the size of

NASA's request moves the program to a 'Grand Probe' rather than a Grand Tour."[35] It was to be less than that.

The reduction in budget for the Grand Tour, an analysis indicated, would likely force delaying the first launch planned for 1975 (to Jupiter, Saturn, and Pluto) to 1977. A 1977 launch, however, would require a more sophisticated TOPS-type spacecraft, which meant additional funding for the Thermoelectric Outer Planet Spacecraft development. It also would require that a seven-segment Titan launch vehicle be available for a 1978 backup. A five-segment Titan rocket, as originally scheduled, could not accomplish the job because of the altered trajectory. Thus, the net result of any delay in the Grand Tour launch date would be to reduce the assurance that the missions could be accomplished and to make it necessary to use an "augmented" and more costly Titan launch vehicle—near-term savings, therefore, would be countered by long-term cost inflation. Reductions in Apollo missions and future human space flights were weighed against Grand Tour, and NASA debated closing down the latter program. What would be the implications of canceling it? If NASA were unable to utilize the "rare and highly productive Grand Tour mission opportunities," administrators reasoned, all would not be lost. The TOPS design provided technology for improving all future spacecraft, and mission design and planning could be used for future investigations of Jupiter and Saturn.[36]

NASA examined the options and decided that Grand Tour could not survive budget realities. Convinced in good measure by his advisors, William E. Lilly and Willis Harlow Shapley, to keep NASA's money focused on human space flight, on December 22, 1971, Administrator James C. Fletcher wrote to Caspar W. Weinberger, deputy director of OMB, and advised him that the TOPS–Grand Tour mission would be terminated. At the time, the cancellation had not been publicly announced, nor had JPL and Caltech been informed, nor had NASA's programmed budget been announced. But the deal was done and public announcements followed. In January, JPL and NASA announced the close of the Grand Tour. The *Los Angeles Times* (January 26, 1972) editorialized, "Dreams of a fantastic voyage faded Monday in the pages of the space budget because of insufficient funds."[37]

Some scientists doubted that there would be an outer planets exploratory mission within the century, for the next opportunity for a Grand Tour would be about the year 2150. This rare alignment of the planets not only allowed a visit to all of the outer planets with only two launches, but because of the slingshot effect around Jupiter, Saturn, and Uranus, it reduced the approximate travel time from Earth to Neptune from thirty to ten years. In an understatement, JPL's director Pickering called the cancellation "disappointing."[38]

George M. Low, NASA's deputy administrator, said the decision to cancel the Grand Tour was "not catastrophic." He suggested that NASA had alternative planes to launch single-planet missions to the outer planets before 1984, missions that would accomplish the same scientific objectives intended for the Grand Tour. He also denied the immediate connection made by the press and many in the scientific community that the Grand Tour was canceled in order to fund the space shuttle. Those decisions, he said, were made independently. "The simple truth," Low remarked, "is that there wasn't unity among scientists regarding the value of the Grand Tour, and this made it an easy target for cancellation."[39]

Bruce Murray concurred. Several weeks before the decision to cancel Grand Tour had been announced, he had written to William Fowler, chairman of the Physical Sciences Committee of the Space Program Advisory Council, that with regard to the Grand Tour, "the efforts of scientists and engineers and NASA managers alike to cooperatively plan an exciting and generally accepted set of flyby missions to the outer planets has proceeded with a good deal of soul searching and criticism from all sides."[40]

Murray reviewed the inception of the Grand Tour, relating it singularly to the TOPS concept, which, he said, was not "a simple extrapolation from previous kinds of space exploration" but something new, and admittedly technologically radical and costly. He thought that "hard headed engineering" and budget realities were bringing the outer planets mission concept back to something on the order of an advanced Mariner approach—something he confessed may have been the correct procedure all along.[41]

That, however, was not the issue. The real issue was that while TOPS and Grand Tour were under development, the Space Science Board and its related scientific groups had become very disenchanted with the Mars Viking program because of rising costs and questionable scientific returns.

The board and a summer study sponsored by the board had narrowly avoided a recommendation to eliminate Viking. The board's hostility was shifted to TOPS and, by association, the Grand Tour. That attack "continued with a long and rather sad display of negativism and unabashed self-interest," Murray suggested, just at the time when NASA and JPL and interested scientists were attempting to restructure and preserve the Grand Tour. The Space Science Board, Murray said, attacked the TOPS and Grand Tour on the grounds they were too big and too costly and would not do enough—thus undermining any attempt to devise an effective program using less costly alternatives, such as Mariner. Murray thought that decision-making authority for the Grand Tour science and engineering should be returned to the project level (i.e., JPL), that NASA should clearly enunciate and defend the goals of the outer planets missions, and that NASA and space scientists must accept and understand the implications of using an advanced Mariner-class spacecraft for the Grand Tour.[42] But, seemingly, it was too late for that.

The Grand Tour of the outer planets was dead. Or was it? A funny thing did indeed happen on the way to Jupiter, Saturn, Uranus, and Neptune.

METAMORPHOSIS

The *Los Angeles Herald Examiner* termed cancellation of the Grand Tour a "sad thing" and noted that the scientific community viewed it as at the least, "unfortunate" and, in the extreme, "disastrous." Science writer David Rubashkin asked, "Who Killed Grand Tour?" He suggested that NASA's perpetual budgetary woes, conflicts, competition among scientists and NASA centers, and competition with the human space program, as well as the war in Vietnam, inflation, and myriad other ailments spelled its doom. But pronouncements of the Grand Tour's death, as it turned out, were premature. "We were not," Ray Heacock affirmed, "going to let it die!"[1]

Those sentiments ran deeply within the Mariner–TOPS–Grand Tour–JPL community. Grand Tour project manager Bud Schurmeier believed that planetary exploration had come too far to be so easily set aside. "Planetary exploration," he later explained, "began as a high-risk enterprise, with relatively limited objectives, in a scientific field characterized by vast unknowns, many speculations, and relatively little public or technical understanding. Launch and flight failures were common. Now a solid foundation of operational experience and ability has accumulated, and techniques have been invented and improved to a high level of sophistication."[2] In the ten years since the Mariner 4 mission to Mars, he said, planetary exploration had reached a "prime level of effectiveness."

"At this moment in the history of planetary exploration," Schurmeier noted, "we stand poised between ability and opportunity."[3] Would NASA and the nation, given that ability, seize the opportunity? There were many who would make it so, including, for example, Schurmeier and Heacock, project manager and deputy manager, respectively, for the Grand Tour; Bill Shipley and Ron Draper, project manager and spacecraft systems manager for TOPS; and Roger Bourke, supervisor for JPL's Advanced Projects Group, who led the mission design work for the Grand Tour. William Fawcett, who had worked on mission science projects for Grand Tour, was determined to see that work through. JPL's response to the alleged death of Grand Tour was more than an "advocacy thing," it was a "cultural thing." The laboratory's scientists and engineers were convinced that they had the needed ability, and they would not let Grand Tour die quietly in 1971. They seized the moment and transformed a demise into an opportunity. "Within ten days time," Draper said, "we countered with a Mariner Jupiter-Saturn Mission."[4]

Draper had been making the Grand Tour "presentation" to NASA and prospective contractors for five years. Costs, he believed, were the determining factor in its closure, and costs were even now threatening the Viking lander program for Mars. James S. Martin at Langley Research Center headed that effort, and Henry W. Norris at JPL served as orbiter manager. Viking had come into being because the Mars Voyager program had become too costly.[5] Could a modified Grand Tour be salvaged in similar fashion?

John Casani, who had worked under Norris as a deputy spacecraft manager on the Mariner 6 and 7 spacecraft, felt that the Voyager Mars project had been "ambitious beyond the capability or technology to do it." JPL "lost control" of the Mars Voyager program to Langley Research Center, and subsequently the program was dropped. Meanwhile, projected Viking costs approached $1.2 billion and Grand Tour costs had grown to $750 million. There were some inherent conflicts in both programs—indeed, the conflicts were at the heart of both manned and unmanned exploration.[6]

Space was a high-risk venture and returns were at best uncertain. Even more at issue in the Grand Tour missions than in the Viking Mars missions was the simple question, as Casani put it, Could you build something that would last that long? Could you build a sophisticated, very advanced machine such as had never been built before, one that would function effectively for ten or twelve years without maintenance and repair? "Scientists were concerned. Congress was concerned," Casani said. What JPL did, he suggested, was scale Grand Tour back to a more believable Mariner Jupiter-Saturn mission for 1977 (MJS '77). Draper was even more succinct in his analysis: "Grand Tour cost $750 million. We countered with a Mariner-J-S mission for $250 million."[7]

To be sure, those numbers had not been easily arrived at, nor were they very firm. Heacock said that in the brief time between the cancellation of Grand Tour and the presentation of a proposal to NASA, JPL put together a number of cost estimates, variously using "system contractors" and JPL managers. Costs for a systems contracting mode, they decided quickly, were too great to be accepted by NASA, so JPL devised a proposal for NASA based on three different cost estimates. The lowest estimate for a flight to Jupiter and Saturn would provide minimal scientific experiments and redundancy features; the medium estimate increased the redundancy features, including better communications and computer guidance systems, and higher costs; the highest estimate provided more security and better science but cost in the $2.25 to $2.75 billion range, which was clearly was unacceptable. The high-end program, however, provided a useful basis for evaluating the risks and returns from less costly exploration ventures. Heacock recalled that NASA discretely "picked something on the low end."[8]

Robert S. Kraemer, who had headed the Advanced Programs and Technology office, replaced his boss, Donald P. Hearth, as head of NASA's Planetary Science Division in December 1971. Kraemer began his space career at an early age. When he was twelve he had become enamored of the idea of space flight. He knew that when he grew up he wanted to build rockets—and the closest he could come to rocketry at American universities in the 1940s was aeronautics. Kraemer enrolled at Notre Dame and received a bachelor's degree in aeronautical engineering; he then transferred to Caltech, where he completed his master's degree in 1951. He joined Rocketdyne, which was then working on ICBM missiles and the J2 rocket engine. He rose through the ranks to become advanced projects manager for Rocketdyne, and after a short time with Convair on a top-priority defense effort to develop an ICBM, moved to NASA for what he expected to be a two- or three-year project. It turned into most of the rest of his working life. He became involved with the Mars Voyager program, and then went to the Planetary Science Division office, where he became a champion of the Grand Tour. He was very surprised when Fletcher canceled Grand Tour, he recalled, and so was Caspar Weinberger in OMB.[9]

Kraemer had scheduled surgery to fuse a disc and was literally "down in the back" when Grand Tour was canceled. The initiative in developing the MJS option, he said, came from Bud Schurmeier and his "gang" at JPL. Kraemer encouraged, incited, and in some respects instigated the JPL Mariner Jupiter-Saturn proposal. He and Schurmeier worked closely on framing the mission, and then he was a key element in putting the proposal on the NASA table. Basically, the idea was to combine the best of the Mariner with the best of the Viking communications and science elements and to borrow as much as possible from TOPS and Grand Tour systems engineering and mission planning. The plan was to launch two Mariner-class spacecraft on a Titan-Centaur vehicle in 1977 for Jupiter and Saturn flyby missions, thus MJS '77. The spacecraft would draw power from radioisotope thermoelectric generators that were designed for the Grand Tour, and each, in the prototype design, could carry more than one hundred pounds of scientific equipment and television cameras.[10]

Kraemer was convinced such a mission would "fly" technically, financially, and politically, and would receive approval of the science community. As a precondition of NASA approval, JPL first secured Space Science Board support for the MJS mission. In addition, Kraemer had strong support among his program heads, particularly Newton Cunningham, who had managed the Mariner 69 (Mariner 6 and 7) program. Caspar Weinberger in OMB was also very supportive, he recalled. Indeed, OMB, which actually had been a supporter of the Grand Tour, had anticipated that NASA would absorb its budget cuts by eliminating the last two Apollo missions, not by eliminating the Grand Tour. Kraemer brought the MJS proposal to John Naugle, associate administrator of the Office of Space Science and Applications. Naugle approved, and with the go-ahead from

NASA administrator James Fletcher, by late February 1972 Naugle had the Mariner Jupiter-Saturn proposal before Congress. Appearing before the Subcommittee on Space Science and Applications of the House of Representatives Committee on Science and Aeronautics, he explained that Grand Tour had been dropped from the program for space exploration because of budget constraints and "the lack of whole-hearted scientific support for the Grand Tour within those constraints." But now, he said, a significant part of the Grand Tour objectives, including flybys of Jupiter and Saturn, and contacts with several of Jupiter's largest satellites and with Saturn's rings and its giant moon, Titan, would be achieved at a far lower cost. While the technical details were still being defined, indications were that the MJS missions would cost less than $360 million, whereas Grand Tour budget requests to Congress for 1972 were in excess of $900 million.[11]

The trip from Earth to Jupiter would take slightly more than one and a half years, to Saturn three and a half years. JPL was to manage the project. Almost as an aside, and certainly ambiguous, was the proviso that "the mission plan will afford flexibility to select a different flyby trajectory at Saturn for the second spacecraft based on data radioed from the first."[12] What did that mean? At the moment, to be sure, no one really knew.

However, in the hearts and minds of its JPL advocates, even in 1972 there was more to MJS '77 than met the eye, or was contained in the proposal or in pronouncements about the proposal. There was what Ron Draper called the "x" factor. The idea was fairly broad, but simple; on this project, one should do nothing that might preclude a longer mission. The "x" factor meant that although the focus on the MJS mission would be Jupiter and Saturn, as well as Saturn's Moon, Titan, the mission would be designed so that if the first mission was successful, the second could be directed on to Uranus and Neptune. Thus, when Roger Bourke began to work on the early MJS trajectories, he found one that could provide the needed "redundancy," that is, that would enable the flight operators to redirect the spacecraft beyond Saturn and on to Uranus. Heacock said that what he and others began to do almost immediately was to "re-engineer our proposal. It was in that context that we made Voyager into what it was to be."[13]

That reengineering took time. "We started designing the spacecraft for Jupiter and Saturn," Schurmeier recalled. "We thought we would follow that with a completely separate Jupiter-Uranus-Neptune project for a 1979 launch, . . . but we were wrong."[14] To paraphrase Bruce Murray, a lot of funny things happened along the way to Neptune. There was a metamorphosis in the works from TOPS, to Grand Tour, to MJS '77, to something yet different and something still some years ahead—a Voyager Grand Tour to the outer planets.

Upon NASA's approval of the MJS '77 project, JPL also began to reengineer its own organization. Schurmeier continued as project manager for the Mariner Jupiter-Saturn 1977 mission. Born in St. Paul, Minnesota, in 1924, Schurmeier was a graduate of New Trier High School in Winnetka, Illinois. He became a naval aviator during World War II and after that enrolled in Caltech, where he received his bachelor's degree in 1945 in mechanical engineering with an aeronautical engineering option. Then he earned his master's degree in aeronautical engineering at Caltech in 1948, received a professional degree in aeronautical engineering in 1949, and in that year joined the Jet Propulsion Laboratory.[15] Schurmeier's task as MJS project manager and deputy assistant director of JPL was to assemble a team to do the job. Perhaps no one knew rocketry and JPL's talents better than he. Part of the job was to fold Grand Tour and TOPS people into the new MJS organization, and part of the job was to leverage people from their work on Viking, Mariner, or Caltech teaching into the MJS program.

Schurmeier asked Ray Heacock, his spacecraft systems manager on the Grand Tour, to continue that work for MJS. There were, however, some serious unresolved management and financial issues relating to the MJS program. Because of the growing budget crises and the very real shortage of personnel through attrition and RIFs (reductions in force), Pickering hoped to utilize contractors for much of the spacecraft integration work while using JPL's personnel for management roles. Pickering, Naugle, Kraemer, and George Low, NASA's deputy administrator, hashed the problem out during the spring, summer, and early fall of 1972. Although JPL had actually begun organizing for MJS '77 in January of 1972, NASA did not issue a task order for the project until May. NASA released a project approval document in June and signed the official MJS project plan in December. It was decided, again, because contracting for the subsystem management on the spacecraft would add $20 million or more to mission costs, that JPL would function both as project manager and as subsystem manager for the spacecraft—which meant, as in the past, that most of the design and integration work would be done by JPL engineers.[16]

Shipley, who had done design work on the Mars Voyager and then was the Flight Projects Office manager for the TOPS program from its inception, joined the MJS team as spacecraft development manager working with Heacock. One of those rare native-born citizens of the District of Columbia, Shipley earned his bachelor's degree in physics from George Washington University and pursued graduate studies under a Saunder's Memorial Fellowship before going to JPL in 1955. Shipley has confirmed that the baseline spacecraft originally proposed for the MJS mission was not the craft that finally flew. Rather, the spacecraft that became Voyager included many TOPS features such as the radio and attitude control systems—and many more.[17]

Ron Draper came in as the spacecraft engineer on MJS from TOPS along with Schurmeier, Heacock, Shipley, and others. Draper stressed that MJS was conceived as a three-and-a-half-year mission, not a ten-year or longer outer-planets mission; it just grew and matured. In time, he and other engineers and technicians began to deal with the spacecraft as though it were a person, not a machine, as it grew into something different than it was first designed to be. Basically, he explained, what they had to do was to scale back from Grand Tour, "to do more, better, and cheaper; to use less power, less money, and less mass." At the time, the longest operating planetary mission had been for nine months, and the most advanced RTGs, used to convert the heat produced from the natural radioactive decay of plutonium, lost power after ten thousand hours, slightly more than one year. Thus, even a three-year mission was a technological challenge. The power problem was eventually solved with the development of special wrappings for the thermocouples that substantially extended the life of the RTGs, and the computer systems were redesigned in order to reduce power consumption.[18]

The spacecraft radio and communications system proved particularly troublesome, and a silent spacecraft is a lost spacecraft. How could one maintain transmissions and reliability for three or more years? In fact, it took three or more years of work on Earth to perfect the antennae and radio transmission and receiver systems for the spacecraft. Ford Aerospace and Philco Radio, working independently and sometimes cooperatively, designed a lightweight spacecraft antenna to receive and transmit communications to and from the Deep Space Network on Earth over distances of hundreds of millions of miles. No one, Shipley noted, had ever built a lightweight graphite-epoxy antenna that size before. Existing antennas were probably a quarter or a third the size planned for MJS, so it was decided to construct five antennas because of the great likelihood of damage. During fabrication, the first casting stuck to the mold and broke, but as it turned out, it could be patched easily, so the order for the four additional antennas was canceled. Traveling wave tubes, a form of low-energy power amplifiers built to operate at the short S- and X-band radio wave lengths, were finally perfected for the MJS spacecraft by the Watkins-Johnson Company. However, the technology involved seemed so questionable that engineers added one solid-state amplifier as a backup. "We took the very best knowledge available to do our work," Draper commented. Nevertheless, as with TOPS and Grand Tour, MJS '77 spacecraft design was an adventure in technology.[19]

The initial design package for the Mariner Jupiter-Saturn spacecraft was completed in July 1972 and revised intermittently thereafter. The initial specifications were simply that the spacecraft should be capable of conducting a Jupiter-Saturn flyby mission and a four-year flight. The spacecraft was to use Viking Orbiter and Mariner designs and "inherited" hardware wherever possible, unless new designs were more cost effective or were absolutely required to complete the unique mission assigned to MJS '77. The initial design specifically excluded the development of radiation-resistant (hardened) parts and circuits, which proved to be a serious oversight corrected only in the final stages of development, and the computers were not to be designed for the continuous storage of data. Spacecraft electrical power was to be provided by three "standard" multihundred-watt radioisotope thermoelectric generators (which turned out not to be standard). The spacecraft was to be "injected" into its trajectory by a five-segment Titan IIIE–Centaur D1-T–Burner 2 (2300) launch vehicle with a 14-foot-diameter shroud covering the spacecraft. Launch system development was under the management of Lewis Research Center (later the John Glenn Research Center), which had oversight for the Centaur. Science, spacecraft design and development, and mission flight design were all under the management of JPL.[20]

Basic phases or sequences in the life of the MJS '77 spacecraft included a launch phase, which took in all of the pre-launch check-outs, the countdown, and the ascent, as well as the navigational fixes—first on the Sun and then on the star Canopus—and the initial science maneuvers. During the cruise phase most systems and science experiments would be quiescent except for navigational tracking. The trajectory correction maneuver phase occurred when course corrections were ordered from the Mission Control Center at JPL. During the encounter phase the spacecraft navigation and controls focused on the best maneuvers and orientation for the scheduled science studies, and on the acquisition of science data.[21]

Engineering elements in spacecraft design had to do with telemetry and guidance systems, data management, command systems and codes, power, structure or mass, navigation, electrical design, mechanical design, environment (the location of sensitive components such as the RTGs and magnetometer), and reliability. Engineering subsections requiring independent team assignments and responsibility included the following:

Structure	Pyrotechnics
Cabling	Trajectory Correction
Propulsion	Temperature Control
Mechanical Devices	Data Storage
Antenna	Radio Frequency
Modulation/Demodulation	Power
Computer Command	Flight Data
Attitude	Control[22]

The MJS spacecraft was to be the product of many diverse technologies and very sophisticated systems engineering and management—and the technologies and the systems were constantly evolving.

Although building a spacecraft for a voyage lasting three or more years was an adventure in technology and engineering, of no less importance in the completion of a successful mission was the initial mission planning, launch vehicle design and construction, integration of the spacecraft with the launch systems, the launch itself, mission and flight management, and, especially, the science—selecting and devising the science experiments and extracting, analyzing and interpreting the data, which is what the mission was all about.

Each of these processes drew upon previous experiences. Roger Bourke, for example, a Stanford University aerospace engineer graduate in 1964, who had collaborated with Gary Flandro in the early analysis of the outer planets Grand Tour opportunity, became supervisor of JPL's Advanced Projects Group and led the early trajectory design work for the Grand Tour. Bourke's design team devised an initial trajectory that would enable the Mariner Jupiter-Saturn spacecraft to fly by Jupiter, Saturn, and Titan, and then continue its mission beyond Saturn. The idea was, in both spacecraft development and trajectory design, to do nothing that would preclude a longer mission. Bourke worked closely with Warren Keller in NASA headquarters, who constantly pressed JPL to do all that could be done with MJS '77, that is, to build and design trajectories to take the spacecraft to as many planets and to visit as many satellites of those planets as possible. Keller wanted numbers: How many planets? How many satellites? How many photographs would be taken? "On the one hand," Bourke recalled, "I felt we were being dragged into more than was prudent and even reasonable to commit to. On the other hand, looking back, no one at the time could have possibly imagined that rich cornucopia of scientific information that came out of these missions. . . . This really was a one-in-a-lifetime opportunity."[23]

The preliminary MJS mission designs were just that, preliminary. For example, the actual flight trajectories were not those previously envisioned or those developed by earlier Grand Tour study groups. Charles Emile Kohlhase Jr., who joined the MJS team in December 1974, became one of the most important managers in the effort. Kohlhase had developed an interest in the planets and a proficiency in math at an early age. When he was about ten, he became a science fiction buff and a fan of writers such as Arthur C. Clarke and Isaac Asimov; he looked at the stars and dreamed about space flight. Born in Knoxville, Tennessee, he moved to Georgia, where he attended Georgia Tech and got his undergraduate degree in physics, despite his father's insistence that he avoid a science major. As an undergraduate he taught mathematics and physics to help pay for his education. After graduation he received a commission in the Navy and served as the electrical officer of a nuclear weapons team aboard the USS *Essex* and *Independence* in the Pacific and in the Atlantic. Then, in May 1959, he joined JPL. He played a leading role in the Mars Voyager and Mariner 6 and 7 mission designs, headed up the navigation development team for the Viking program, and later joined the MJS '77 team navigation systems group as mission analysis and engineering manager.[24]

Kohlhase was the mission architect of what would become the Voyager Grand Tour of the outer planets. His role was to orchestrate the interplay of requirements and plans for how the mission, flight, and ground systems would perform together to achieve the maximum scientific value return while limiting the level of risk to the completion of the mission. Mission planning would ensure that nothing in the prime mission design for Jupiter and Saturn would preclude a flight beyond Saturn. However, the immediate concern was simply getting a spacecraft to Jupiter and Saturn, and to the vicinity of as many of their satellites as possible. Selecting the right trajectory was a most complex, complicated business, for Earth and each of the planets are in constant motion about the Sun and their satellites, in turn, are in motion about each planet. A launch from Earth to Jupiter could occur every thirteen months, and the MJS launch was targeted for launch periods in 1976, 1977, or 1978. During any of those years there was roughly a thirty-day interval when a launch could be made to Jupiter because of the position of that planet relative to Earth. On the day selected for launch, because of Earth's rotation, there was approximately a one-hour "firing window."[25] The problem was to select the year, day, and hour for launch, as well as the trajectory, so that the spacecraft would achieve optimum flybys of the planets to perform the science missions. In addition, one—and ideally both—MJS spacecraft should encounter Jupiter's satellite Io and the other three large moons, if possible, and then proceed to Saturn and pass as close as practicable to its rings and great moon, Titan.

It was, Kohlhase said, a fascinating business with an almost infinite number of variables. "Trajectory people," he noted, "basically generate iso-energy curves—a 'pork-chop curve,' which produce a 'quite extended' number of arrival dates." The navigation team first generated 10,000 possible trajectories, then proceeded to analyze those so as to best match encounters with mission objectives. That process resulted in the selection of the best 100–110 trajectories. Then, those trajectories were subjected to what Kohlhase called the "nudge factor," that is, course corrections or changes during the flight. External conditions and launch constraints, including development of the spacecraft and launch vehicles, actual delivery dates, the selection of the science packages, propellent consumption, communications requirements (could data be transmitted to Earth during critical encounter periods?), and unforeseen events, also affected the final trajectory selection.[26]

One unforeseen event was that Pioneer 10, in December 1973 the first spacecraft to fly through the asteroid belt and fly by Jupiter, reported very intense radiation belts around the planet, posing a much greater hazard than scientists had predicted. Pioneer 11, which passed Jupiter in 1974, confirmed the findings. These discoveries resulted in eliminating some of the MJS flight options, but, more significantly, the MJS spacecraft design had to be changed and "hardened" to prevent radiation damage. For example, tantalum shielding was added to many components. Thomas Gavin, the project assurance manager on the MJS, a graduate of Villanova University in chemistry and a veteran of the Mariner program, took the lead in redoing the electrical systems so they would be resistant to radiation. "Jim Briden worried about the components, Tom Gindorf worried about the shielding, and I worried about the circuits," he recalled. Thus, whereas the original baseline fabrication design for the MJS spacecraft specifically excluded hardening for radiation, the version that finally flew, "because of the radiation problems, [included] . . . extensive margins in the electronic design, so that we knew the spacecraft could last a long time."[27]

Richard Laeser, an MIT electrical engineer, came to JPL in 1964 after serving a tour in the Army Signal Corps. He earned the master's degree in electrical engineering from the University of Southern California, became Deep Space Network manager for a number of the Mariner flights, and about 1974 was asked by Bud Schurmeier to join the MJS team as mission operations system manager. That meant, Laeser explained, that he had responsibility for the design of the ground system, the control center and the plans and procedures to be used during flight. After the launch, he was the mission director during the Jupiter and Saturn encounters.[28]

His most lasting impressions have to do with how a group of discordant people, personalities, and procedures began working together as a team. There was at first a lot of stumbling and some mistakes, but the team became "wonderfully efficient and enthusiastic" and usually accomplished its goals. Another recollection of the pre-launch phase involved the tension between the spacecraft designers and the operations designers: "I still have the image of the continuing struggle—push-pull back and forth—between the spacecraft designers and the operations designers, and the feeling of always losing, being on the operations side, to the decisions made by the spacecraft designers." The spacecraft engineers pushed the state of the art on the spacecraft, and the operations people were supposed to figure out how to fix up any problems later.[29] Indeed, there were problems later, and indeed they were fixed.

What the proposed MJS mission was all about was the scientific investigation of the outer planets, which, of course, had been the purpose of the Grand Tour. Grand Tour science and Grand Tour scientists were the essential ingredients of the theoretically more limited Mariner Jupiter-Saturn project. However, Grand Tour science, as well as the Grand Tour program, went through an MJS metamorphosis and emerged from the Jupiter-Saturn chrysalis to reemerge as a "Voyager, Grand Tour, outer-planets, science, exploratory adventure."

Edward Carroll Stone Jr. became a key player in MJS and subsequently Voyager science. Born in Knoxville, Tennessee, and raised in Burlington, Iowa, on the upper reaches of the Mississippi River, Stone was strongly influenced by his high school physics instructor, Wilford White. White, and the dawn of the atomic age, whetted Stone's interest in nuclear physics. He enrolled at the University of Chicago and left there with a doctorate in physics in 1964. In 1959 at Chicago he began working with John Simpson and participated in the design of experiments for Earth satellites. Then, in 1964, Rochus Vogt, a friend of Simpson's, asked Stone to come to Caltech, where the former was engaged in high-altitude balloon experiments that resulted in the discovery of cosmic ray electrons. During those early years Stone had little connection with JPL. However, in 1970, Vogt, who was a member of the study team for the Grand Tour, asked him to join that group, which was primarily concerned with scientific instrumentation and power requirements, and which essentially defined the scientific character of the Grand Tour.[30]

When Grand Tour was canceled, Stone's association with outer plan-

Edward C. Stone. (NASA photo P-39689b)

ets exploration faded—for a short time. Upon the resurrection of MJS '77, and following discussions with Bud Schurmeier, and at the urging of Caltech's provost, Stone agreed to accept the job of project scientist. At that point the broad character, if not the specifics, of the science experiments were already defined. Grand Tour science became MJS '77 science.[31]

MJS science initiatives really began in 1970, with NASA's invitation for proposals for Grand Tour science experiments. In April 1971 John Naugle in NASA's Office of Space Science selected a team of 108 scientists from those who had submitted proposals to define the Grand Tour of the outer planets scientific mission. Those picked were organized into thirteen different scientific teams, but that effort seemingly died with the demise of Grand Tour. Then it too, like Grand Tour, was resurrected at least in part. In December 1972 JPL selected ninety scientists from the Grand Tour "pool" to comprise an MJS '77 Science Steering Committee. They were now organized into eleven, rather than thirteen areas as had been the case under the Grand Tour, and each group, other than the Imaging and Radio

Science Teams, whose equipment was specified and provided by NASA headquarters, was to be responsible for the design and construction of the instruments associated with its area of investigation. Each team was headed by a principal investigator who reported to Ed Stone, who, in turn, reported to his counterpart in NASA's Planetary Science office, Milton A. Mitz, the program scientist, and also to JPL's Project Manager, Bud Schurmeier.[32]

The two MJS spacecraft were to have identical instrumentation but different trajectories. Their overall mission was to visually characterize Jupiter, Saturn, their satellites, and the rings of Saturn; to study the atmospheres (if any) and sizes of ring particles and the masses of the planets and satellites using radio science; to examine their atmospheres and surfaces with infrared spectroscopy and radiometry; and to analyze the upper atmospheres and ionospheres of the planets and their satellites with ultraviolet spectroscopy. Science experiments included studies of magnetic fields, plasma (the properties of electrons and positive ions in the solar wind and within the "magnetospheres" of the planets), cosmic rays, radiation belts, interplanetary particulate matter, and other properties of the planets, the rings, and the satellites using photometry, polarimetry, and radio astronomy.[33]

Members of the Imaging Team and the Radio Science Team selected by NASA included the following:[34]

IMAGING TEAM

Bradford A. Smith, New Mexico State University, team leader
Alan F. Cook, Smithsonian Astrophysical Observatory
George E. Danielson Jr., Jet Propulsion Laboratory
Merton E. Davies, Rand Corporation
Garry E. Hunt, Meteorological Office, England
Tobias Owen, State University of New York
Carl Sagan, Cornell University
Lawrence A. Soderblom, United States Geological Survey
Vernon Suomi, University of Wisconsin

RADIO SCIENCE TEAM

Von R. Eshleman, Stanford University, team leader
John D. Anderson, Jet Propulsion Laboratory
Thomas Croft, Stanford University
Gunnar Fjeldbo, Jet Propulsion Laboratory
George S. Levy, Jet Propulsion Laboratory
G. Leonard Tyler, Stanford University

The principal investigators who headed up the other science teams included the following:

INFRARED SPECTROSCOPY AND RADIOMETRY

Rudolf A. Hanel, Goddard Space Flight Center

ULTRAVIOLET SPECTROSCOPY

A. Lyle Broadfoot, Kitt Peak National Observatory

MAGNETOMETRY

Norman F. Ness, Goddard Space Flight Center

PLASMA

Herbert S. Bridge, Massachusetts Institute of Technology

LOW ENERGY CHARGED PARTICLES

Stamatios M. (Tom) Krimigis, Johns Hopkins University

INTERSTELLAR COSMIC RAYS AND PLANETARY MAGNETOSPHERES

Rochus E. Vogt, California Institute of Technology

INTERPLANETARY/INTERSTELLAR PARTICULATE MATTER

Robert K. Soberman, Drexel University/General Electric

PHOTOPOLARIMETRY

Charles F. Lillie, University of Colorado

PLANETARY RADIO ASTRONOMY

James W. Warwick, University of Colorado

By late 1972, when Stone joined the MJS team, many of the engineering-related science decisions had already been made, but most of the mission details, science sequences, and many conflict resolutions lay ahead. The process of selecting experiments for the flight began with a non-NASA peer review of the proposals. Then, "in-house" NASA science panels rated those selections under what Stone termed loosely categories A, B, C, and D. A meant the proposal was ready to go; B was "okay, but"; C was a possible; and D meant "forget it." However, Stone's job had as much to do with integrating the science proposals with spacecraft engineering work as with the pure science, for he was the "impedance matcher" between scientists and engineers. A good part of his task had to do with getting the two groups to ask the right questions of each other.[35]

As Kohlhase explained things, it was a matter of how to combine needs with reality. For example, there were often conflicting needs among the scientists, who, in the selection of trajectories, established their preferences as to where and how their instruments should be pointed at the time of encounter. Kohlhase and Stone established "value zones," which helped the scientists make sometimes painful selections. Stone's great talent, Kohlhase thought, was in being able to create a consensus among the scientists and then helping to identify and incorporate the required engineering into the final decisions.[36]

In September 1973, nine months after the initial science selections, Stone supervised a confirmation review where some of the experiments were removed and substitutions were made. For example, an interstellar wind experiment was added but later deleted. Issues that arose at that point related to data coding, data rates, and transmission channels. Another issue was whether to use reaction wheels on the spacecraft to give the operators more precise control over camera pointing; the consensus was that this was a good idea, but additional costs and development time were too great. Then, too, how long should the boom holding the magnetometer and other instruments be? There were the persistent questions of engineering needs versus science returns, such as how "clean" the radioactive fuel should be, for that decision could affect the operation of a number of the science instruments. Perhaps the biggest issue was just where the science experiments should be located on the spacecraft to make them most effective while containing costs and considering the limited design options.[37] All in all, there were a lot of quid pro quos in juggling the needs of engineering and science preferences.

There was an even more fundamental constraint affecting both science and engineering: funds were limited, and the funding that had been approved ended as of June 1980, although mission planning and science experiments were based on four years of operation.[38]

By early 1973, even a four-year MJS mission seemed doubtful, for on January 5, NASA announced a new round of program reductions to "adjust its activities in space and aeronautics to a lower spending level." The agency was to remove seven hundred civil servants from its payrolls. The space shuttle, scheduled for its first orbital flight in 1978, was to be delayed, and a high-energy astronomy observatory was shelved. NASA planned to retain the Skylab experimental space station program just begun, the Apollo-Soyuz Project, Viking, the space shuttle, and the Mariner Jupiter-Saturn mission.[39] Funding for all NASA programs, however, was less assured than in the past. Mariner voyages to Jupiter and Saturn appeared most vulnerable to budget problems and were beset by competing scientific programs. Even as late as 1974, the metamorphosis of Mariner Jupiter-Saturn into a Voyager Grand Tour remained an unlikely prospect.

5

MARINER JUPITER-SATURN '77

Grand Tour was canceled. In its place, Mariner Jupiter-Saturn '77 came into being, a program providing severely limited planetary exploration as compared to Grand Tour and proffered by NASA with lower expectations and at considerably lower cost. The primary objectives of the new project were to "extend the exploration of the solar system to the neighborhoods of Jupiter and Saturn with a spacecraft that can conduct significant scientific experiments . . . and pave the way for later missions to the outer planets." But a funny thing also happened to MJS '77, for it became something other than that which it was supposed to be. Indeed, the transformation began almost at the point of inception of the program, for the Mariner Jupiter-Saturn spacecraft was to be more than "an extension of the Mariner spacecraft" family as announced in the Jet Propulsion Laboratory's *Annual Report* for 1973–74. Instead, it was to be a reengineered, redesigned, re-created Mariner, which, though it drew heavily from earlier Mariner and Viking spacecraft designs, had become, by launch time, something distinctively different.[1]

It was a Mariner that was, in fact, greatly "augmented in capability," as JPL reported, "to meet the requirements for long life, long-range communications, precision navigation, solar-independent power, and for support of the science investigations." The reengineering was predicated, Ray Heacock explained, on the proposition that nothing should be done to compromise the possibility that the spacecraft might fly beyond Jupiter and Saturn, say, to Uranus and Neptune. Indeed, not only was nothing done to prevent that from happening, but much was done to assure that it could happen. This meant that Grand Tour was alive, if not well. Thus began the transformation of MJS '77 into what became Voyager Grand Tour; a quiet, subtle transformation, almost unrecognized, even by its architects.[2]

To be sure, the JPL/NASA horoscope failed to offer auspicious signs for a Grand Tour anywhere, or even a less-than-grand tour to Jupiter and Saturn. On January 3, 1973, within months of approving the MJS program, NASA announced that it was "starting *today* to make a number of program reductions to adjust its activities in space and aeronautics to a lower spending level." Fiscal constraints and program reductions became the order of the day.[3]

In spite of dark economic clouds on the NASA horizon, JPL's Mariner Jupiter-Saturn team did not despair but, paradoxically, sensed opportunity. When JPL got the formal go-ahead for the project, Ray Heacock and other engineers restudied the project schedule and realized that as the work involved primarily retooling the Mariner and Viking spacecraft, the five-year interval to the launch period was more than ample time. "We set about to take advantage of the fact that NASA had given us good budget line items going into 1973 and 1974" and for the initial startup in 1972, Heacock said. "So we set about to re-engineer our proposal."[4] During the reengineering, which lasted through 1973 and much of 1974, JPL engineers transformed the MJS program into something other than what it was scheduled to be.

The special mission requirements for trips to Jupiter and Saturn (and beyond) included a power source independent of solar energy because of the greater distances from the Sun; markedly improved communications systems because of the vastly greater distances than those experienced in Moon, Mars, or Venus missions; improved attitude-control systems with three axes, stabilization reference points that usually included the Sun,

Earth, and a star such as Canopus; and a distinctive science payload integrated into the spacecraft design and into the mission plan. Over twenty-five changes were incorporated as alterations to the basic Mariner design. Modifications included "a new Radio Frequency Subsystem design, a new Flight Data Subsystem, a new digital computer-based Attitude and Articulation Control Subsystem, the integration of the final kick stage rocket into the Propulsion Module as part of the spacecraft, a new electronic equipment packaging approach, and increase in the science payload to 105 kg and 100 W."[5]

The MJS mission required a spacecraft that could make its own navigational corrections, but also would be responsive to Earth controls; that is, it needed autonomous, self-repairing, self-correcting computers that could interface with mission control. In effect, it needed to have the qualities of a self-test-and-repair computer, but without the weight, power requirements, and costs of a STAR. The MJS Engineering Design Team decided to use a Viking computer (available at a nominal cost) with the addition of a new plated-wire memory backup system and two computer command subsystems to handle flight data and attitude control as well as to conduct routine checks on power, radio frequency signals, errors, and system failures.[6] The MJS was given two computer command subsystems with two processors and two memory systems, unlike most previous spacecraft, which used only one, or occasionally two, independent computer subsystems, for the new ones were interlinked rather than independent, and were uniquely designed to work together. About half of the memory of each computer was protected and used only for storage of fixed routines; the remainder was used in tandem with its opposite number for spacecraft sequencing or navigation. The basic plan was to update the computers' programming (sequence load) every four weeks during the cruise phase of the flight. Then, during a planetary encounter phase, more frequent reprogramming from ground control was anticipated that could alter the spacecraft's maneuvers, or change the position or operation of its instruments, recorders, or data formats.[7]

A flight data subsystem (FDS) and a data storage subsystem were specially designed for the MJS mission. Data were categorized into four basic types: (1) engineering only, (2) cruise science and engineering, (3) encounter general science and engineering, and (4) imaging plus general science and engineering. Data was to be handled at rates ranging from 115,200 bits per second (bps) for imaging to 10 bps for engineering. Cruise science data rates ranged from 2,560 bps near Earth to 80 bps at Saturn, while the data storage subsystem equipped with a digital tape recorder could store 5×10^8 (500 million) bits.[8]

As Heacock noted, "earlier Mariner and Viking Orbiter missions to Mars achieved maximum information rates of 16,200 bps." However, the more complex science payloads scheduled for missions to Jupiter and Saturn required a substantial increase in the rate of data handling, which was a significant challenge to spacecraft system designers. Although JPL's Deep Space Network was composed of state-of-the-art equipment, the far greater distances involved in flights to Jupiter and Saturn as contrasted to Venus and Mars meant correspondingly reduced received signal power. The decision was made to use two S-band transmitters, which received at about 2,110 MHZ and transmitted at 2,300 MHZ, and two X-band transmitters, which operated at about 8,400 MHZ. However, the higher frequency of X-band also created a greater potential for loss or dispersion of the signal, and should the X-band signal be lost due to the to fact that the transmitters were new and less tested, it was decided as a backup to enhance S-band signals by using a so-called Reed-Solomon encoder rather than the more conventional Golay encoder.[9]

This was all a bit complicated, but it sounded simple as explained by Heacock—and it was also rather far-reaching in that the radio data transmission systems, like other components of the MJS spacecraft, were being designed for something beyond Jupiter and Saturn:

> We said if we lose both X-band transmitters and have to go with the S-band, then we can enhance the return of the data with the Reed-Solomon encoder. The Golay encoding was being used routinely in the early part of the mission. . . . You have one encoding bit for every bit you're sending so you're losing 50 percent of the capacity of your system just to add the error detection and correction. The other reason we built the data system was so that you could run both processors simultaneously but independently so that one could routinely handle the data formatting while the other one was doing data compression on the imaging data, the high-rate data. So this was another way of building in a capability in the system to enhance the data return if we were to

lose X-band, or if we were to be able to go on to Uranus. *Going to Uranus and Neptune was an option we always kept in our minds.*[10]

The electrical power consumed by radio communications, on-board computers, and science packages had to come from other than solar-power sources, for at great distances from the Sun such as those of Jupiter and Saturn, not to mention Uranus and Neptune, there was little solar energy to collect. So MJS design engineers, including Ron Draper and Bill Shipley, who had been respectively the design team manager and the project manager for that project, turned to their TOPS design experiences for a solution. As previously mentioned, that concept involved the use of radioisotope thermoelectric generators. Although TOPS had been a design project only, it had established the need for such a power device, and coupled with Air Force interest in a similar power system (for classified communication satellites being developed by Lincoln Laboratory at Massachusetts Institute of Technology), RTG development by the Department of Energy's Energy Research and Development Administration (ERDA), and its primary contractor General Electric, matched the time frame for the MJS spacecraft launches.[11]

The incorporation of advanced technology as represented in the MJS spacecraft design was more than a matter of convenience or luck, let alone selecting from the existing stock in the warehouse of technology. Rather, the RTG, like the Mariner, interplanetary rocket-booster systems, and the Deep Space Network and interplanetary communications systems, was an innovation and an adaptation of all parts of the JPL engineers' collective memory and experience. Although the technology existed, for the most part it had never been assembled together or used in the format designed by JPL engineers for the MJS spacecraft. Thus, taken all in all, the MJS spacecraft "became new technology."

That applied to the propulsion systems as well. The proposed launch vehicle, a two-stage Titan rocket (Titan IIIE) built by Martin Marietta Corporation with solid-rocket booster motors and an upper-stage liquid oxygen–liquid hydrogen Centaur rocket, could not produce the velocity needed to achieve the desired trajectory. However, the Centaur was particularly suitable for interplanetary flights because its engines had a high specific impulse and could be restarted after a long shutdown period.

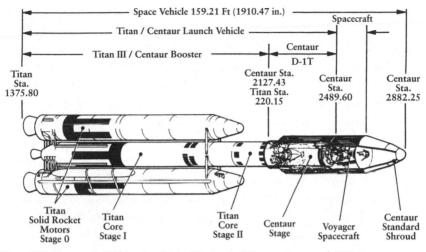

Titan IIIE Centaur D-1T launch vehicle. (Heacock, "Voyager Spacecraft," 214)

Thus, the design team decided to use a solid-rocket "kick" motor to be constructed by Centaur's developers (General Dynamics and Convair) under the management of NASA's Lewis Research Center, and as a new innovation decided to incorporate the kick motor as a part of the spacecraft. Engineers at JPL designed a propulsion module within the spacecraft to include the TE 364 motor and, by doing so, reduced the overall weight below that required by an independent kick rocket—a solution that increased the available science payload and reduced costs.[12] It was a good piece of design engineering, and, if other adaptations had not already done so, it converted what had ostensibly been a Mariner spacecraft into something new.

There were more innovations in the propulsion systems. Instead of using highly pressurized, cold nitrogen gas as the fuel for the attitude-control jets, as on previous Mariner and Viking spacecraft, Heacock's design team, working with Lewis engineers, decided to convert to low-pressure, hot hydrazine fuel, and to combine all the fuel used for trajectory correction maneuvers, thrust vector control, and attitude controls into a single tank, which was a novel approach. By doing so, the MJS spacecraft fuel tank was less susceptible to explosion and there could be better in-flight management of fuel reserves.[13]

Charles Kohlhase became the "keeper of the hydrazine." The secret to maintaining adequate fuel to allow for unforeseen flight corrections meant, for the most part, careful mission planning that would, as much as possible, prevent shortages. In turn, that meant planning the initial launch such that the planetary explorer reached Jupiter at a time when it could fly by the Jovian moon Io, an essential science objective. Io was in proper position every six days, which meant that mission trajectories had to conform to arrival dates that occurred at six-day intervals. Another critical target, Saturn's giant satellite Titan, was in a position for a flyby every sixteen days, and thus the MJS launch time and trajectories had to be consistent with both the six- and sixteen-day intervals.

Moreover, Saturn's equatorial rings and the possibility of extended material outside of the main rings interfering with the spacecraft, dictated a mission plan that would bring the ship closest to the planet below the equatorial plane. In other words, the flight had to be aimed to dip "underneath" Saturn.[14] And, of course, with all of this, there was the need to reserve as much fuel as possible to exercise additional options, such as an extended cruise to Uranus and Neptune.

There was yet much more to mission planning and fuel conservation, for each planned trajectory required a fuel consumption analysis, and every modification required an adjusted trajectory and thus another fuel evaluation. Then, too, reoutfitting the MJS spacecraft with tantalum shielding to resist the electron radiation discovered at Jupiter by Pioneers 10 and 11 added mass and meant recalculating trajectories and fuel usage. Voyager 1, with its heavier titanium shielding, was rerouted to fly closer to Jupiter in order to get better science, while Voyager 2 was to give the planet a wider berth in order to protect the spacecraft for a possible extended mission.[15]

"What we were doing," Kohlhase said, "was constantly discovering and defining the constraints or potential problems, then running the trajectories available, and then evaluating the available propellent." Kohlhase had come from the Viking program to head the MJS Mission Analysis and Engineering Team in December 1974, when Ralph F. Miles decided to leave. It was, Kohlhase recalled, the job he had always dreamed of—"the best job in the world"—and it kept him fully occupied for the next fifteen years.[16]

Planetary exploration was a matter of constant planning, continually choosing among options and alternatives, and, for the most part, taking calculated risks. It was also a matter of teamwork and good systems engineering. MJS '77 was a team effort structured under a matrix organization with the project management drawing when and as needed from virtually the entire resources of the Jet Propulsion Laboratory, as well as from industry through the contract processes on an when-and-as-needed basis. In general, the contracting and fabrication mode began in 1975 after the completion of the spacecraft design, science plan, mission design, navigation plan, launch vehicle design, operations program, and general mission profile.[17] The MJS '77 flight probes were designed, assembled, and tested by JPL.

As spacecraft manager working under project manager Bud Schurmeier, Heacock headed the JPL "matrix" team whose members in turn interfaced with the various divisions and JPL project offices. Bill Shipley, called the development manager, was Heacock's deputy. Ron Draper, the systems engineer, headed the overall engineering design work from the inception of MJS '77 until the vehicles were ready for launch at Cape Canaveral, while William Fawcett was the instrument manager. Tom Gindorf headed the Environmental Requirements Team, which was concerned not only with decontamination of the spacecraft for its interplanetary voyages, but also, along with Thomas Gavin, the quality and reliability assurance manager, he assessed the impact of the space environment on mission operations. D. D. Howard headed the vital Quality Assurance Team, which was responsible for testing components and for interfacing components with each other as well as with the science packages and the overall spacecraft. Joseph Shaffer Jr. and George W. Haddock served at different times as the MJS launch vehicle interface managers, who worked with the Lewis Research Center's Centaur team and with General Dynamics contractors on the problems of interfacing the launch vehicle, that is, the Titan, Centaur, and "kick" rocket stages, with the overall spacecraft itself and overall mission design.[18]

Like the tip of a technological iceberg, the spacecraft design team represented only the exposed or more visible part of a systems engineering structure that permeated JPL and extended far beyond. Subsystem elements—ranging from the launch operations at Kennedy Space Flight Center to launch vehicle development at Lewis Research Center and science experiments based variously at Goddard Space Flight Center and univer-

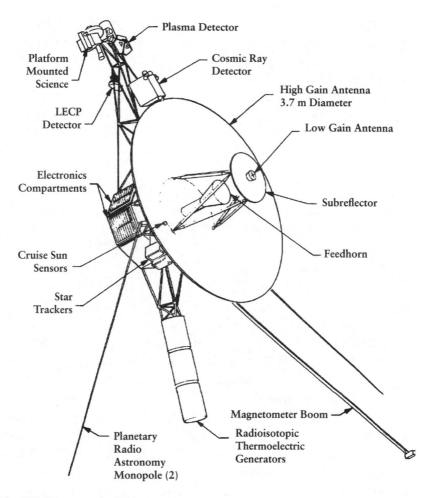

Platform Mounted Science
Plasma Detector
Cosmic Ray Detector
High Gain Antenna 3.7 m Diameter
LECP Detector
Low Gain Antenna
Electronics Compartments
Subreflector
Cruise Sun Sensors
Feedhorn
Star Trackers
Magnetometer Boom
Planetary Radio Astronomy Monopole (2)
Radioisotopic Thermoelectric Generators

The MJS '77 spacecraft, 1974.

sities across the nation, not to mention hundreds of contractors and sub-contractors—were integrated into the MJS design and development effort.

At NASA Headquarters, the Office of Space Science and Applications, headed by John Naugle through June 1974 and then by Noel W. Hinners, had oversight of the MJS '77 program. The Lunar and Planetary Programs Division, headed by Robert Kraemer when the program came into being, was responsible for the actual administration of the program until 1976, when he turned over his duties to A. Thomas Young. Rodney A. Mills was

MJS '77 program manager for the Lunar and Planetary Division and along with his deputy program manager, Arthur Reetz Jr., and the program scientist, Milton A. Mitz (to whom JPL's chief project scientist, Ed Stone, reported) were the key Headquarters supervisors and contacts with JPL.[19] For the most part, in fitting with JPL's expertise, its culture, and its past relations with NASA, Headquarters granted JPL virtual autonomy in the technical and engineering management of the MJS '77 program.

Communications and interaction were the keys to the success of such a large-scale technical enterprise. For example, Schurmeier held regularly scheduled monthly meetings with all project heads and JPL division representatives and scheduled special "all-hands" meetings as required. Draper and Kohlhase held regular staff meetings where the broadest possible issues and problems were addressed. Representatives from each of the project's working groups attended the weekly staff meetings and anyone related to the project had access to these meetings and to Draper and Kohlhase. When problems arose requiring some expertise from outside the MJS management framework, as they often did, the systems manager's job was to see that such was included within that framework and that the interface between them was effective.[20]

Building a successful interplanetary spacecraft was a matter of anticipating problems before they occurred and recognizing and solving problems when they did occur—all the way down to launch. One of the first such problems had to do with the command computer subsystems derived from the Viking program. Designers modified the Viking CCS to better fit MJS mission requirements, and then, in order to save money, tacked on the modification to the existing Viking contract with General Electric. The "improvement" designed for the MJS mission resulted in the CCS circuit boards being "over-etched," causing the boards to fail to make connections between their various layers. It cost about $3 million for JPL and General Electric engineers to perform the rigorous review needed to detect and solve the problem.[21]

Over-engineering itself could be problem. For example, as mentioned earlier in passing, one group of JPL engineers thought it would be necessary to use reaction wheels on the MJS spacecraft to provide the pointing control and stability needed for scientific observations. Others, including Ray Heacock, thought reaction wheels were unnecessary and expensive, and that stability and control could be maintained with attitude-control thrusters and the basic inertial reference unit comprising three gyros (gyro A: pitch and yaw, gyro B: roll and pitch, gyro C: yaw and roll). In the end, the decision was made not to include the reaction wheels, thus saving $500,000, and avoiding the addition of more complicated systems.[22]

Another crisis involved the RTGs. A material used as insulation sublimed, that is, it passed from a solid to a gaseous state and then redeposited itself on an insulating blanket, causing an electrical short in the system and lowering the power output to less than the required levels. The solution, discovered by an RCA engineer, was to put a nitride coat on the thermocouple device, which prevented the subliming and the shorts.[23]

One of the last "Heacock hardware headaches," as the problems came to be known, was the apparent failure of an integrated circuit after the spacecraft had been shipped to Cape Kennedy, only weeks before the scheduled launch. An internal short in the integrated circuit of the CCS developed when one of the persons fabricating the circuit board dipped hot parts into a beaker of water to cool them down. This caused pin holes in the circuits and created a migrating gold resistive short, a form of dendritic growth promoted by the water vapor in the air and inside the chip, which combined to create leaflike conducting crystals. Those were, Heacock recalled, "desperate times."[24] Problems, as it turned out, by no means ended with the launch.

In October 1974 Schurmeier conducted a comprehensive Mission and Systems Design Review that generated 116 "concerns" or potential problems. He categorized them into ten major "action areas" and required that each be addressed by the appropriate project group and that a report or resolution of the problem was to be presented at the Final Spacecraft System Design Review scheduled for March 1975. There was widespread concern because the computer command subsystem, the flight data subsystem, and the attitude and articulation control system (AACS) relied on having surplus or redundant memory in their shared memory system. How would the failure of any one of those systems and its memory banks affect the operation of the others?[25] Problems abounded.

How would bad weather affect X-band radio signals during and after launch? Would there be any scheduling conflicts with the Deep Space Network that would prohibit the transmission or receipt of signals to or from the spacecraft during encounter periods? How would environmental requirements, such as radiation hardening for the Jupiter encounters and shielding from solid particles in Saturn's outer rings, affect the performance of the spacecraft? Then too, reaction-wheel versus non-reaction-wheel configuration still needed a final review.[26]

The mission plan required that all commands to the spacecraft be sent twice in order to assure correct reception, but could the transponder simply feed the commands through in a loop? What if the signals were simply regurgitated by the spacecraft? Or what would happen to signals when the spacecraft was not Earth-oriented? These were concerns for review by the Data Systems Team. Did the navigation plans need sharper confirmation and definition? Were there potential flaws in the optical navigational system? How did the science and navigational radio command requirements match up? What trade-offs were going to be necessary? There were numerous unresolved questions relating to the processing and delivery of data being returned from the spacecraft; in particular, could the costs of imaging frames be reduced? How would data collected during encounters be handled by the ground data system? What schedules had been developed for the delivery of experimental data records and supplementary experimental data?[27] All of these issues had to be addressed by March 1975, in time for the final design review.

There was yet another concern flagged by John Casani, then chief of Division 34, JPL's Guidance and Control Division. There was, he noted on the standard form Mariner Jupiter/Saturn 1977—Concern/Action Control Sheet, "no plan for sending a message to our extra solar system neighbors." What did he recommend? "Send a Message!"[28]

The result of that memo turned out to be one of the most publicized aspects of the entire interplanetary mission and a public relations coup for NASA. Pioneer 10, launched for a Jupiter flyby in March 1972, and Pioneer 11, launched in April 1973, were both sent on trajectories that would, assuming the spacecraft survived, ultimately take them outside of the Solar System decades after radio contact with Earth had ceased. Both carried an engraved plaque of the human physique and greetings to extraterrestrials, for the idea of life on Mars or on planets of other suns had tugged at the heartstrings of humanity throughout most of the twentieth

Table 5.1

Sounds of Earth of Voyager (in Sequential Order)

Whales	Footsteps and heartbeats	Horse and cart
The Planets (music)	Horse and Carriage	Volcanoes
Laughter	Train whistle	Mud pots
Fire	Tractor	Rain
Tools	Truck	Surf
Dogs, domestic	Auto gears	Crickets, frogs
Herding sheep	Jet	Birds
Blacksmith shop	Lift-off of Saturn 5	Hyena
Sawing	Rocket	Elephant
Tractor	Kiss	Chimpanzee
Riveter	Baby	Wild dog
Morse code	Life signs—EEG, EKG	Ships
Pulsar		

century. The concern that MJS '77 should carry appropriate greetings was sufficiently important that NASA sought the counsel of American astronomer Carl Sagan of Cornell University. Sagan and his wife Linda had helped radio astronomer Frank Drake devise the Pioneer message. One of the nation's most widely read astronomers, Sagan was editor of the respected journal *Icarus* and prominent in the field of planetary studies.[29]

He chaired a committee of astronomers and scientists who devised an exceptional "Greetings to the Universe." Committee members included Frank Drake, Cornell University; A. G. W. Cameron, Harvard University; Phillip Morrison, MIT; Bernard Oliver, Hewlett-Packard Corporation; Leslie Orgel, Salk Institute; Alan Lomax, Choreometrics Project, Columbia University; Robert Brown, Center for World Music, Berkeley, California; Murry Sidlin, National Symphony Orchestra, Washington, D.C.; and artist Jon Lomberg of Toronto, Canada. The committee decided that the spacecraft should carry representative sounds and images on a 12-inch gold-plated copper disk in an electronic format, a message that would portray the character and diversity of life on Earth.[30]

Given that decision, what *does* one select to best represent life on Earth? Sagan and his associates selected 115 images and sounds from Earth, including surf, wind, and thunder; bird calls; and the sound of whales and other animals (see tables 5.1 and 5.2). The recording also included ninety

Table 5.2

Voyager Record Photograph Index

---

1. Calibration circle, Jon Lomberg
2. Solar location map, Dr. Frank Drake
3. Mathematical definitions, Dr. Frank Drake
4. Physical unit definitions, Dr. Frank Drake
5. Solar system parameters, Dr. Frank Drake
6. Solar system parameters, Dr. Frank Drake
7. The Sun, Hale Observatories
8. Solar spectrum, H. Ecklemann
9. Mercury, NASA
10. Mars, NASA
11. Jupiter, NASA
12. Earth, NASA
13. Egypt, Red Sea, Sinai Peninsula, and the Nile, NASA
14. Chemical definitions, Dr. Frank Drake
15. DNA Structure, Jon Lomberg
16. DNA structure magnified, Jon Lomberg
17. Cells and cell division, Turtox/Cambosco
18. Anatomy 1, Field Enterprises Educational Corp and Row, Peterson and Co.
19. Anatomy 2, Field Enterprises Educational Corp and Row, Peterson and Co.
20. Anatomy 3, Field Enterprises Educational Corp and Row, Peterson and Co.
21. Anatomy 4, Field Enterprises Educational Corp and Row, Peterson and Co.
22. Anatomy 5, Field Enterprises Educational Corp and Row, Peterson and Co.
23. Anatomy 6, Field Enterprises Educational Corp and Row, Peterson and Co.
24. Anatomy 7, Field Enterprises Educational Corp and Row, Peterson and Co.
25. Anatomy 8, Field Enterprises Educational Corp and Row, Peterson and Co.
25a. Human sex organs, *Life: Cells, Organisms, Populations*
26. Diagram of conception, Jon Lomberg
27. Conception, Lennart Nilsson
28. Fertilized ovum, Lennart Nilsson
29. Fetus diagram, Jon Lomberg
30. Fetus, Dr. Frank Allan
31. Diagram of male and female, Jon Lomberg
32. Birth, Wayne Miller
33. Nursing mother, UN
34. Father and daughter (Malasia), David Harvey
35. Group of children, Ruby Mera
36. Diagram of family ages, Jon Lomberg
37. Family portrait, Nina Leen
38. Diagram of continental drift, Jon Lomberg
39. Structure of earth, Jon Lomberg
40. Heron Island (Great Barrier Reef of Australia), Dr. Jay M. Pasachoff
41. Seashore, Dick Smith

Continued

Table 5.2 Continued

42. Snake River and Grand Tetons, Ansel Adams
43. Sand dunes, George Mobley
44. Monument valley
45. Forest scene with mushrooms, Bruce Dale
46. Leaf, Arthur Herrick
47. Fallen leaves, Jodi Cobb
48. Sequoia, Josef Muench
48. *[sic]* Snowflake, R. Sisson
49. Tree with daffodils, *Gardens of Winterthur*
50. Flying insect with flowers, *Borne on the Wind*
51. Diagram of vertebrate evolution, Jon Lomberg
52. Seashell (Xancidae)
53. Dolphins, Thomas Nebbia
54. School of fish, David Doubilet
55. Tree toad, Dave Wickstrom
56. Crocodile, Peter Beard
57. Eagle, Donona
58. Waterhold, South African Tourist Corporation
59. Jane Goodall and chimps, Vanne Morris-Goodall
60. Sketch of bushmen, Jon Lomberg
61. Bushmen hunters, R. Farbman
62. Man from Guatamala, UN
63. Dancer from Bali, Donna Grosvenor
64. Andean girls, Joseph Scherschel
65. Thailand craftsman, Dean Conger
66. Elephant, Peter Kunstadter
67. Old man with beard and glasses (Turkey), Jonathon Blair
68. Old man with dog and flowers, Bruce Baumann
69. Mountain climber, Gaston Rebuffat
70. Cathy Rigby, Philip Leonian
71. Sprinters (Valeri Borzov of the USSR in lead), *The History of the Olympics*
72. Schoolroom, UN
73. Children with globe
74. Cotton harvest, Howell Walker
75. Grape picker, David Moore
76. Supermarket, H. Eckelmann
77. Underwater scene with diver and fish, Jerry Greenberg
78. Fishing boat with nets, UN
79. Cooking fish, *Cooking of Spain and Portugal*
80. Chinese dinner part, Michael Rougier
81. Demonstration of licking, eating, and drinking, H. Eckelmann
82. Great Wall of China, H. Edward Kim
83. House construction (African), UN

Table 5.2 Continued

84. Construction scene (Amish country), William Albert Allard
85. House (Africa), UN
86. House (New England), Robert Sisson
87. Modern house (Cloudcroft, New Mexico), Dr. Frank Drake88. House interior with artist and fire, Jim Amos
89. Taj Mahal, David Carroll
90. English city (Oxford), *C. S. Lewis, Images of His World*
91. Boston, Ted Spiegel
92. UN Building Day, UN
93. UN Building Night, UN
94. Sydney Opera House, Mike Long
95. Artisan with drill, Frank Hewlett
96. Factory interior, Fred Ward
97. Museum, David Cupp
98. X-ray of hand, H. Eckelmann
99. Woman with microscope, UN
100. Street scene, Asia (Pakistan), UN
101. Rush hour traffic, India, UN
102. Modern highway (Ithaca), H. Eckelmann
103. Golden Gate Bridge, Ansel Adams
104. Train, Gordon Gahan
105. Airplane in flight, Dr. Frank Drake
106. Airport (Toronto), George Hunter
107. Antarctic Expedition, *Great Adventures with the National Geographic*
108. Radio telescope (Westerbork, Netherlands), James Blair
109. Radio telescope (Arecibo), H. Eckelmann
110. Page of book (Newton, *System of the World*)
111. Astronaut in space, NASA
112. Titan Centaur Launch, NASA
113. Sunset with birds, David Harvey
114. Spring Quartet (Quartetto Italiano), Phillips Recordings
115. Violin with music score (Cavatina)

minutes of music from different cultures and eras (see table 5.3), including Eastern and Western classics (Bach, Beethoven, Mozart, and Stravinsky, as well as Javanese gamelan, Indian raga, and Chinese ch'in), ethnic music, and "music from Senegal, Australia, Peru, Bulgaria, and Azerbaijan, as well as rock and roll."[31]

A major portion of the recordings were the spoken greetings from the people of Earth in fifty-five languages, including Akkadian, the most ancient known language, spoken in Akkad and Sumer some six thousand

Table 5.3

Music on Voyager Phonograph Record (in Sequential Order)

---

1. Bach Brandenburg Concerto no. 2, first movement, Karl Richter conducting the Munich Bach Orchestra.
2. "Kinds of Flowers," Javanese court gamelan. Recorded in Java by Robert Brown, Nonesuch Explorer Records.
3. Senegalese percussion. Recorded by Charles Duvelle.
4. "Pygmy Girls Initiation Song." Recorded by Colin Turnbull (Zaire).
5. Australian Horn and Totem song. Recorded in Australia by Sandra LeBrun Holmes. Barnumbirr–Morning Star Records.
6. "El Cascabel." Performed by Lorenzo Barcelata and the Mariachi Mexico.
7. "Johnny B. Goode." Performed by Chuck Berry.
8. "New Guinea Men's House." Recorded by Robert MacLennan.
9. "Depicting the Cranes in Their Nest." Recorded by Coro Yamaguchi (Shakubachi).
10. Bach Partita no. 3 for violin, Gavotte en Rondeau, Arthur Grumiaux, violin.
11. "Queen of the Night" (aria no. 14), from Mozart's *Magic Flute,* Edda Moser, soprano.
12. "Chakrulo." Georgian (USSR) folk chorus.
13. Peruvian pan pipes. Performed by José María Arguedas.
14. "Melancholy Blues." Performed by Louis Armstrong. Columbia Records.
15. Azerbaijan. Two flutes. Recorded by Radio Moscow.
16. *Rite of Spring,* conclusion, by Stravinsky. Igor Stravinsky conducting the Columbia Symphony Orchestra.
17. Bach Prelude and Fugue no. 1 in C Major from the *Well Tempered Clavier,* book 2. Glenn Gould, piano.
18. Beethoven's Fifth Symphony, first movement. Otto Klemperer conducting. Angel Recording.
19. "Izlele Delyo hajdutin," Bulgarian shepherdess song, sung by Valya Balkanska.
20. Navajo Indian Night Chant. Recorded by Williard Rhodes.
21. "The Fairie Round," from Pavans, Galliards, Almains. Recorded by David Munrow.
22. Melanesian pan pipes. From the collection of the Solomon Islands Broadcasting Service.
23. Peruvian song. Recorded in Peru by John Cohen.
24. "Flowing Streams," Chinese ch'in music. Performed by Kuan P'ing-Hu.
25. "Jaat Kahan Ho," Indian raga. Performed by Surshri Kesar Bai Kerkar.
26. "Dark Was the Night." Performed by Blind Willie Johnson.
27. Beethoven String Quartet no. 13, *Cavatina.* Performed by the Budapest String Quartet.

---

years ago (see table 5.4). All of Earth's major languages and some of the more exotic were used; the greetings concluded with a phrase in Wu, a modern Chinese dialect. Sagan captured the spirit of the interplanetary exploring adventure with the observation that "the spacecraft will be encountered and the record played only if there are advanced spacefaring

Table 5.4

Languages Heard on Voyager Record (Not in Sequential Order)

Sumerian	Spanish	Turkish	Swedish
Akkadian	Indonesian	Welsh	Ukrainian
Hittite	Kechua	Italian	Persian
Hebrew	Dutch	Nguni	Serbian
Aramaic	German	Sotho	Luganada
English	Bengali	Wu	Amoy (Min dialect)
Portuguese	Urdu	Korean	Marathi
Cantonese	Hindi	Armenian	Kannada
Russian	Vietnamese	Polish	Teluga
Thai	Greek	Mandarin	Hungarian
Romanian	Latin	Gujorati	Czech
French	Japanese	Ila (Zambia)	Rajasthani
Burmese	Punjabi	Nyanja	

civilizations in interstellar space. But the launching of this bottle into the cosmic ocean says something very hopeful about life on this planet."[32]

President Jimmy Carter's greeting on the disk was equally eloquent:

> We cast this message into the cosmos. It is likely to survive a billion years into our future, when our civilization is profoundly altered and the surface of the Earth may be vastly changed. . . . This is a present from a small distant world, a token of our sounds, our science, our images, our music, our thoughts and our feelings. We are attempting to survive our time so we may live into yours. We hope someday, having solved the problems we face, to join a community of galactic civilizations. This record represents our hope and our determination, and our good will in a vast and awesome universe.[33]

The recordings, produced as a public service by Columbia Records, were packaged in an aluminum jacket, with a cartridge, needle, and instructions in symbolic language on how to use the recording. The idea of encapsulating the history and sounds of life on Earth captured the imagination of the American people and resonated around the globe, if not in the heavens. The selection of languages, sounds, photographs, and music provided a remarkable glimpse of life on Earth that may have had far greater meaning and significance to humans on that Earth than to the

Table 5.5
Mission Phase Definitions

Phase	Period Covered	Source of Definition
Launch	L to MM separation	Launch activity period
Near-Earth cruise	MM separation to celestial acquisition	MM stabilization period
Far-Earth cruise	Acquisition to (~)L + 70d	HGA not pointed at Earth
Earth-Jupiter cruise	L + 70d to J − 90d	HGA cruise-tracking activities
Jupiter encounter	J − 90d to J + 50d	High-data-rate science and navigation activities
Observatory	J − 90d to J − 30d	Non-real-time high-rate data
Far encounter	J − 30d to J − 1; J + 1d to J + 10d	Real-time high-rate data
Near encounter	J − 1d to J + 1d	Near-encounter maximum real-time activities
Post encounter	J + 10d to J + 90d	General science data
Jupiter-Saturn cruise	J + 90d to S − 90d	HGA cruise-tracking activities
Saturn encounter	S − 90d to S + 50d	High-data-rate science and navigation activities
Observatory	S − 90d to S − 30d	Non-real-time high-rate data
Far encounter	S − 30d to S − 1d; S + 1d to S + 10d	Real-time high-rate data
Near encounter	S − 1d to S + 1d	Near-encounter maximum real-time activities
Post encounter	S + 1d to S + 90d	General science data
Post-Saturn cruise	S + 90d to EOM	HGA cruise-tracking activities

Notes: L = launch time; J = Jupiter; S = Saturn; MM = mission module; HGA = high-gain antenna; EOM = end of mission.

prospective spacefaring beings from an extraterrestrial civilization in a galaxy far, far away.

Back on Earth, by the close of 1974 the reengineering design work had been completed. A mission plan that integrated program requirements with the spacecraft's components was approved and then circulated to all JPL divisions on December 2. This plan included a master science plan devised by the Science Steering Group, which was a committee composed of representatives of all the science projects scheduled for MJS '77.[34] The mission phase definition profile approved by the group illustrates the character of the mission and the essential elements of mission planning (see table 5.5).

When Schurmeier held the Final Spacecraft System Design Review in March 1975, actions were completed or underway on all of the concerns previously identified. A mission operations strategy and a mission design review was scheduled for August, with a final selection of flight trajectories expected in January 1976.[35] While the spacecraft design was moving ahead nicely, the mission design was still taking shape, but the transformation of MJS '77 into something other than it was supposed to be was almost complete. The Mariner spacecraft designed by JPL for travel to Jupiter and Saturn was clearly capable of more, and NASA now began to reconsider and resurrect elements of the Grand Tour.

In March 1975, NASA issued a request to prospective contractors for the preliminary design of a planetary probe capable of studying Jupiter, Saturn, Uranus, and Neptune, and compatible with either a Pioneer or a Mariner spacecraft and their respective launch vehicles. "Taking advantage of an outer planet alignment that occurs only once every 180 years," the *Defense/Space Daily* newsletter observed on April 4, "NASA is looking toward the launch of a single spacecraft in 1979 to successively fly by Jupiter, Uranus and Neptune."[36] But NASA also was considering the possibility of adapting the ongoing Mariner Jupiter-Saturn 1977 mission, or the Pioneer Venus 1978 launch, for the extended mission.

Then there was no further news on the subject for a time. However, in May 1975 the Department of Energy approved a contract with General Electric's Space Division at a cost of $17,736,000 for the fabrication of RTGs to be used aboard MJS '77. In June 1975, there were substantive administrative changes at JPL, which, though not directly related to MJS '77 or the Grand Tour, had consequences for both. Bruce Murray, a champion of the original Voyager program and of the subsequent Grand Tour of the outer planets, became director of Caltech's Jet Propulsion Laboratory, succeeding Bill Pickering, who was retiring after twenty-one years as head of the laboratory. Even then, Murray was trying to promote world wide scientific support for a proposed International Solar System Decade that would focus research and space exploration on all aspects of the Sun's family.[37]

Murray, deeply involved in planetary studies, was completing scientific descriptions of Venus and Mercury from Mariner 10 data before his promotion, and was also comparing the surface histories of Mercury, Venus,

Mars, and the Moon with that of Earth. Thanks to Sagan, Murray, Pickering, and many others, planetary studies were very much in the public eye during the 1970s. Moreover, by 1975 there were distinct improvements on the economic outlook. Almost every business index began to show improvement; inflationary pressures and the petroleum and monetary crises triggered by the OPEC oil embargo in mid-1973 eased. In addition, national political crises that had resulted in the resignation of the vice president and, on August 9, 1974, the resignation of President Richard M. Nixon, brought a return of seeming stability. Confidence and public optimism began to return, and the NASA budget situation began to improve. As a result, JPL increased its efforts to obtain NASA approval for an "extended" MJS '77 interplanetary mission. In July 1975, NASA adopted an official logo for the MJS '77 mission and began to study the possibility of retargeting the second Mariner Jupiter-Saturn '77 mission for a Uranus encounter. That was, of course, an option that JPL engineers had constantly before them from the first moment of the inception of the MJS program, an option with roots in the Ranger, Mariner, Mars Voyager, and Grand Tour programs, and in the hearts and minds and history of JPL.

VOYAGER

Prospects for extending the Mariner Jupiter-Saturn missions to include Uranus or Neptune soon dimmed. In 1975 JPL suggested that two additional Mariner missions be scheduled for launch to Uranus and Neptune in 1979. When NASA demurred, program manager Bud Schurmeier suggested modifying the MJS missions by adding one launch targeted for Uranus and Neptune in 1979, providing a scheduled launch in 1977 (Jupiter), 1978 (Saturn), and 1979 (Uranus and beyond), thus providing a better budgetary "spread." The Uranus-Neptune launch would only be made if the prior two Mariner missions were successful.[1] That also failed to win NASA approval and, indeed, by the mid-1970s, NASA's program options had become severely limited.

The space shuttle had become NASA's primary agenda following the close of the manned lunar program with the flight of Apollo 17 in December 1972. Preliminary design studies for a fully reusable and cost-saving orbital space vehicle had begun in 1969, and in July 1971 NASA selected Rocketdyne to build the main engines for the proposed shuttle. North American Rockwell (later Rockwell International) received an interim contract for the construction of the shuttle spacecraft itself in August 1972 and a final contract in April 1973. NASA spent $475 million on shuttle development in 1974, $805 million in 1975, and $1.2 billion in 1976; costs rose steadily to approximately $2 billion annually by 1981, when the shuttle made its first test flight.[2] The space shuttle predetermined NASA programs and policies for the next several decades.

In the face of constricted budgets and severe inflationary pressures, NASA's decision to build the space shuttle assumed the premise that the shuttle would become its primary launch vehicle. Thus, NASA abandoned the Saturn rockets it had developed for the lunar program and as long-term launch vehicles. The shuttle decision also meant that NASA would, as much possible, forego the use of the Titan rockets being procured from the Air Force at considerable cost. Indeed, the MJS '77 missions were predicated on the use of the last two of NASA's already procured Titan IVs. The shuttle decision also meant that NASA had decided to delay or abandon planetary missions in favor of Earth-orbital missions, and that future scientific exploration would focus on Earth rather than on the other planets.[3] Murray, JPL's new director, objected strongly. "He fought like hell for the planetary program to try and keep it alive," Schurmeier observed. There was, Schurmeier thought, an attitude developing that now that the United States had won the race to the Moon, "we'd better start working on problems here on Earth."[4]

Congressional preferences turned from pure science and exploration to practical applications. For example, Senator Edmund Muskie, then a Democrat candidate for the presidency, argued that the nation's priorities had turned from space to "hungry children, inadequate housing, decaying cities, and insecure old age." The shuttle, an Earth-orbital vehicle, was in part a response to those changing priorities, and so were Skylab and the later decision to build the International Space Station. Applications, that is space and science projects relating to Earth resources, communications, navigation, national security, climate, weather, pollution, and agriculture increasingly became the focus of the Office of Space Science and Applications, and of NASA as a whole. The upper-stage Centaur rocket, used with the Atlas for the Mariner 6 and subsequent Mariner launches and for the Pioneer 10 and 11 interplanetary missions, became the vehicle of choice for Earth-science and communication satellite launches.[5]

Despite improving economic conditions, and the enthusiasm, dedication, and foresight of the MJS '77 interplanetary exploration team of engineers and scientists; public, congressional, and NASA support for an extended mission of planetary exploration such as a Grand Tour waned. Even the basic Mariner mission to Jupiter and Saturn confronted diminishing interest and support. With design work completed and fabrication well underway, through 1975 and 1976 the MJS interplanetary exploration mission continued to face many obstacles and uncertainties.

For one thing, the leadership changed. Murray, a geologist, came to JPL from the science side of the Caltech community, and though none exceeded his dedication to planetary exploration, he was not an engineer and was, for the most part, an untested administrator. "His biggest administrative job had been running a campus geology project that employed six persons and a budget of $200,000 a year," one author noted. He was described, moreover, as a "tall, square-jawed man more comfortable giving orders than listening to advice" whose style of leadership tended to be aggressive, if not abrasive.[6] He was one of the uncertainties.

Murray created yet another uncertainty for the MJS program by selecting its mentor and program manager, Bud Schurmeier, to head up a new JPL civilian space applications program consistent with the trend in NASA to seek to apply space technology to Earth-related problems. As Schurmeier explained it, "The United States had won the race to the moon, and the space program was going down some, and the Lab was embarking on non-space activities." He assumed management of the applications activities, and in 1981 became JPL's associate laboratory director for Defense and Civil Programs.[7] His departure from MJS '77, a program that in many ways represented the culmination of his aerospace engineering expertise and his JPL career, left a void.

It was this management gap into which John Casani stepped and soon filled easily, confidently, and competently—after having raised an initial "firestorm." Born and reared in Philadelphia, Pennsylvania, Casani completed Jesuit High School with a strong background in Latin, Greek, English, French, geometry, and classical studies. He entered the University of Pennsylvania as a liberal arts major, and after several years of being unable to select a major, his father thought it might be a good time for him to go into the military for a few years to mature. Casani matured quickly.

John Casani. (NASA photo P-4659)

He talked to his college roommate about options, and the latter suggested he switch to an electrical engineering major. The School of Engineering agreed, if he took some remedial courses, and his father also agreed. His roommate said he would help Casani "catch up" but the next fall never showed up at school—having flunked his freshman engineering. However, Casani stuck with it and graduated in 1955 with a degree in electrical engineering. During those years, he recalled, he read a series of articles on space exploration in *Collier's* magazine written by Wernher von Braun, among others. Already a big science fiction fan, he was captivated by space.[8]

After graduation Casani accepted a job in Rome, New York, working on electronic countermeasures. That winter the temperature never got higher than five degrees below zero. He and a chemical engineering friend kidded about going to sunny, magical California. And they did. Casani "holed up in a fraternity house in Southern California" until he found a job—with JPL. A recruiter, Jack James, sent him a telegram or called him every day. He accepted the job because he was "wanted" and it involved space, although the pay was a good bit less than what was being offered by North American Aviation. At the time, JPL operated under a contract with the Army Ballistic Missile Agency; there was no NASA, and Caltech's presence at the lab was almost unnoticed. His first work was on a radio inertial guidance system for the 1,500-mile Jupiter missile and the last three Corporal missiles used to develop and test the new guidance system. Following the Corporal tests at White Sands, New Mexico, Casani and his team were sent to Cape Canaveral in Florida for a test of the Jupiter. "We were all down at the Cape with the vehicle on the pad" when the decision was made that the Air Force, rather than the Army, was to have responsibility for ICBM (intermediate range) missile launches, and the Jupiter launch was halted.[9] Then NASA came into the picture.

Bill Pickering indelibly imprinted on Casani's mind the seriousness of the changeover to NASA. It was, Pickering explained, "like a girl on the street corner waiting to be picked up. Once she got into that car, she will never be the same. And that is what is going to happen to us. It will be a major change in direction for us. It will be a great ride, but we are going to be different from this point on."[10] Pickering was right.

As Casani moved from Corporal to Pioneer to Ranger to Mariner, the spacecraft designs kept evolving. Mariner 4 and following vehicles had brand new architecture. And then came MJS '77, new and distinctly different from the Mariners that had come before. Soon after Murray became director, Casani received a telephone call from Gen. Charles Terhune, JPL's assistant director, asking him to leave the job he had managing Division 34 and join Schurmeier as deputy on the MJS '77 program with the idea that in three months he would become program manager. Schurmeier was to be moved to head up the new space applications program. Casani wanted to know if his present boss, Jack James, knew of the change. Terhune told him no, James was in South America. Casani thought he should talk to James before he made the decision to change. Terhune told him he needed an answer in twenty-four hours and mentioned, in a nonthreatening way, that the last person to whom we offered a job failed to take it—and that was the last job he had ever been of-fered.[11] But Casani had already decided to take the job.

By 9:00 A.M. the next day Casani was working directly under Bud Schurmeier. Schurmeier was "wonderful," Casani recalled. "He really coached me and tried to make sure that [I understood] everything that was important that went on, such as the interactions with Headquarters and the science community, and every place else." At the time, the MJS team reported to Casani through Ray Heacock, the spacecraft system manager. Under Heacock were Bill Shipley, who managed all of the engineering elec-tronics and hardware, and Bill Fawcett, who handled the science instru-ments. Charles Kohlhase was the mission manager who did all of the tra-jectories and analysis, and Ed Stone was project scientist. At the time Stone was still teaching courses at Caltech and was supposed to be on a one-third assignment with the MJS project, which probably meant, Casani noted, that Stone was spending no less than thirty hours a week on MJS, "because we saw plenty of him." Richard Laeser was on Mission Opera-tions, and Ron Draper was the spacecraft engineer.[12]

Among other things, Schurmeier impressed upon Casani the reality that the MJS '77 program had been approved "by the skin of our teeth. It was close to being killed." The eventual success of the program had to do with the fact that Schurmeier focused just on Jupiter and Saturn. That was the mission and he did not want anyone thinking about anything else, but Casani was not able to fully adjust to that. One of the first things he did when he joined the MJS team was to ask for a new telephone number, specifically for 864-6578 (or 4-MJSU), the MJSU standing for "Mariner Jupiter-Saturn-and-Uranus." Although he approved it, Schurmeier did not like the request because he did not want anyone talking publicly about Uranus.[13]

As Division 34 manager, Casani had been building guidance and con-trol systems for the MJS mission and had instructed his team that they would not deliver any piece of hardware to the MJS project that might place an inherent limitation on how long the mission could last. They built for something "beyond Uranus and Neptune." The fact was, Casani stressed, that Grand Tour and TOPs experiences provided the intellectual, conceptual framework for the MJS '77 mission.[14] And there was something else that may have had a bearing on shaping the MJS missions to become something other than that they were professed to be.

In late 1975 Casani and his brother were on a sailboat heading for Catalina Island. He looked up at the night sky and saw an array of all the visible planets just above the horizon: "There was Mercury, Venus, Mars, Jupiter, and Saturn. You could see them all." They saw them from the boat, with the knowledge that in a few years the alignment would be suitable for a flyby by the MJS spacecraft.[15] It was an inspiring moment.

There had to be more to MJS than a Mariner going to Jupiter and Saturn. Certainly there had to be something to distinguish the Mariner Jupiter-Saturn mission from previous missions launched under such vague titles as Mariner '69, Mariner '73, Mariner Mars, Mariner Venus, Mariner 6, or Mariner 7. Casani, who had been with Mariner since Mariner numbers 1 and Mariner 2, had difficulty distinguishing one mission from the other because of the similarity of their names. The title MJS '77 did not help, for there was more to this spacecraft than a Mariner. It was a distinctive space vehicle, as Mariner 4 had been unique from its predecessors, and its mission, at least in the JPL sense of things, was much more than "MJS." It was designed to do more and it could do much more, but it needed a name befitting its uniqueness, or at least so Casani thought.[16]

He decided that a name change was in order. That created a firestorm, for Casani was the "new guy on the block." The MJS team had just gone through a big contest to select the MJS emblem/logo "with 'MJS 77' all over it," and there was some emotional attachment to the project name. Still, in the early spring of 1976 Casani decided to hold a contest for a new name, with a case of champagne going to the winner. Soon people began seriously thinking not only about the name change but once again about the full potential of the Mariner mission—the idea had struck a chord. The name finally selected, one that stood out clearly over other options, was Voyager.[17] Voyager, but not yet Grand Tour.

There was a yet another anomaly in the Voyager naming, for Voyager 1 was to be launched three or four weeks after the launch of Voyager 2. The nature of their trajectories was such, however, that the second Voyager to be launched would arrive at Jupiter first, get to Saturn first, and lead thereafter. Thus, if the first launch were called Voyager 1, it would only be number one for about three months before being overtaken by the second spacecraft. Kohlhase and Casani threshed out the problem and decided that to avoid a public relations nightmare the spacecraft to be labeled Voyager 1 would be the first to get to the planet, not the first launched.[18]

There was more to the decision to rename the program Voyager, for the JPL/MJS team believed, now more firmly than ever, that Voyager could and should do more than go to Jupiter and Saturn. The idea was that if Voyager 1 completed its science objectives at Jupiter and the primary science experiments scheduled for Saturn, and Voyager 2 made a successful flyby of Jupiter, then the flight path of Voyager 2, which lagged behind Voyager 1 at the Saturn approach by approximately nine months, could be very slightly adjusted to allow Voyager 2 to proceed on to Uranus and Neptune. It had all been worked out, Casani said. "We knew what the strategy was; we knew what we were going to do; we knew what the decision points were," and Kohlhase had the trajectory work to support it.[19] NASA Headquarters became more warmly disposed to the idea.

In February 1976 NASA issued a final project approval document for a mission it now styled "Mariner Jupiter/Saturn 1977 Planetary Exploration (Outer Planets Missions)." NASA had not yet assigned the "Voyager" nomenclature. The objectives of the missions were described as in the past, namely, to conduct exploratory investigations of the Jupiter and Saturn planetary systems as well as the interplanetary medium between Earth and Saturn. This program was to be accomplished by two launches in 1977 on flyby trajectories that would use Jupiter's gravity assist to reach Saturn. However, a new line was added to the mission description, denoting a new possibility that had not been a part of previous MJS '77 mission descriptions: "Should the spacecraft continue to operate past the Saturn encounters, an extended mission would be proposed in anticipation of penetrating the boundary between the solar wind and the interstellar medium to allow measurements to be made of interstellar fields and particles unmodulated by solar plasma. The option of retargeting the second mission for Uranus encounter is under study."[20]

Aviation Week & Space Technology reported that because NASA could not procure an independent budget item for a third Mariner Jupiter-Uranus launch (as had been urged as an option by JPL), it was now considering the possibility of sending one of the Mariner Jupiter-Saturn spacecraft on to Uranus after its Saturn encounter scheduled for 1981. But the report added that "NASA planetary officials consider a Uranus flyby with the second MJS spacecraft as a long shot at best."[21]

By the end of February (which notably closed on the twenty-ninth in the year 1976), however, NASA had approved a Mariner Uranus mission option. It meant, Casani explained to the press, that "if the first spacecraft mission was successful and satisfied scientific requirements" and "if the second Mariner showed continuing signs of a long and active life," the decision would "probably be made to aim the second Mariner so that when it swung back by Saturn's equator it would head off in the direction of Uranus, four more years away." If all went well the Mariner would fly by Uranus between November 1985 and January 1986 and then proceed farther out into the Solar System. And, he added, it would take a "miracle . . . for the Mariner to continue in working order all the way out to Neptune."[22] But it *could* happen. Thus, MJS '77, in February 1977, six months before the scheduled launches, had become Voyager and, just possibly, Voyager Grand Tour.

In March 1977 NASA officially accepted Voyager as the name of the two spacecraft scheduled to be launched that summer "for an extensive reconnaissance of the outer planets." If all went well, one of the Voyagers would be targeted for the first encounter with Uranus, some 1.7 billion miles (2.7 billion kilometers) from Earth, "and possibly Neptune," 2.7 billion miles (4.3 billion kilometers) distant. The first Voyager was expected to encounter Jupiter in March 1979, passing within 222,000 miles of the planet, and would take closeup photographs of Jupiter's four largest moons. It was to arrive at Saturn in November 1980, some 130,000 miles distant from the planet, passing within 4,000 miles of Titan, the planet's largest moon. Now NASA described project Voyager as the "next step" in the U.S. program of systematic planetary exploration. By these explorations of the Solar System, scientists expected to "learn more about the history and future of the solar system and particularly our own planet Earth."[23] The MJS mission concept had expanded to become Voyager, and the Voyager missions held the prospect of reinstating the Grand Tour of the outer planets.

But the road from JPL, where the Voyager spacecraft was designed, assembled, and tested, to Florida for an August launch was a long one, and strewn with obstacles. Texas congressman Jack Brooks, chairman of the House Committee on Government Operations, who had been a stalwart partisan of the manned space program and of the Texas-based Lyndon B. Johnson Space Center, wanted NASA to explain a $35.7 million discrepancy between what it claimed to be the costs of the planned Voyager missions and cost estimates developed by the Government Accounting Office (GAO). The direct costs of the Voyager program, NASA announced, were $320 million, plus project-related costs including launch vehicles, tracking facilities, and flight support estimated at an additional $130.9 million; for a program total of $450 million.[24]

The GAO reported to Congress that NASA had failed to include in $23.6 million in unreimbursed costs paid to the Energy Research and Development Administration for the development of the RTGs for the spacecraft. Neither had NASA included the $5 million cost of Goddard Space Flight Center's civil service positions used in support of the Voyager program, nor $3.8 million contributed by NASA's Low Cost System Office toward development of hydrazine thrusters and inertial navigation systems. In addition, NASA spent another $3.3 million to improve infrared imaging equipment.[25] In an age of tight budgeting, strict accountability, and waning congressional and public support for space science, the GAO report cast a cloud over Voyager.

In response, NASA noted that the RTGs were useful in all Department of Defense and NASA space missions, not just Voyager. It did not have the responsibility or the ability, it said, to monitor costs incurred by other agencies using their own funds. Moreover, NASA regarded the Goddard Civil Service salary costs as fixed costs which had no direct applicability to one project but were part of NASA's baseline capability. And NASA had, as required, submitted its Project Status Report to the committee having oversight for its activities (the Senate Committee on Aeronautical and Space Sciences) and "would be glad to submit them to any other congressional committee requesting them."[26] NASA's response to the GAO report was sent to Congressman Brooks in early August, only weeks before the first scheduled Voyager launch.

The two Voyager spacecraft were already on site and being readied for launch. The first to reach Kennedy Space Center at Cape Canaveral in Florida was a nonflight, proof-test model of the Voyager used to examine spacecraft components and study methods of inspection, replacement, and repair. That spacecraft (VGR77-1) was packaged and shipped to Florida as a pathfinder on a practice drill in March 1977, preceding the shipment of the actual flight-ready spacecraft.[27]

Jet Propulsion Laboratory completed a preshipment review of its two Voyager flight-ready spacecraft in April 1977. Identified as VGR77-2 and VGR77-3, they were almost—but not quite—identical. The total system included the mission module and the propulsion module; the spacecraft itself weighed 1,820 pounds, including the 231-pound science instrument payload; the integral last-stage rocket weighed 2,690 pounds; and the total launch weight of the vehicle with the adaptor joining the spacecraft to the Centaur stage was 4,630 pounds. These spacecraft were hybrid Mariners that used thermoelectric power rather than solar cells, possessed an improved communications systems using the largest communications antennae ever flown on NASA space missions, and were more automatic and independent of Earth-based control than any previously launched spacecraft. Three RTGs provided the electrical power, and each Voyager contained three engineering subsystems programmable for on-board control of spacecraft functions, including the computer command subsystem, the flight data subsystem, and the attitude and articulation control subsystem. Each subsystem had an independent but interactive memory bank with a combined data storage capacity of 536 million bits. These subsystems could be updated or modified by ground controls, and ground control alone was able to make trajectory corrections.[28]

Following check-out, JPL engineers loaded the spacecraft and accessories aboard special air-ride trucks. The first caravan of trucks left Pasadena on April 21 and the second on May 19. After they arrived in Florida the spacecraft were subjected to intense systems checks and problems were encountered in the VGR77-2 spacecraft, the one scheduled for the first launch. The attitude-control flight software was not ready. Shorts were discovered in the integrated circuits of the computer command subsystem, and another potential problem developed in the flight data subsystem. On one occasion, Ray Heacock recalled, Vince Edwards looked in and announced that the traveling wave tubes for radio transmission were failing. The failure turned out to be in the test console, not in the spacecraft systems.[29] It was good that the Voyagers were twin spacecraft.

One difference between the two probes was the higher power available from the RTGs of VGR77-2, because it had been prepared for a continuation to Uranus. Engineers removed the RTGs from that spacecraft and placed them in the twin VGR77-3, which would now become the first Voyager to be launched (and thus Voyager 2). Despite a six- or seven-day delay in the review process, the replacement enabled NASA to hold to the original launch date while the problems in VGR77-2 were assessed and corrected. Another problem was solved by taking the AACS from the test model and putting it in place of the original VGR77-2 unit.[30] Barring other unforeseen circumstances, NASA expected to launch Voyager 2 on August 20, and Voyager 1 on September 1.

At that point the Voyager Engineering Design Team had finished their work; Mission Operations would soon take over and implement the work of the mission design team. The mission profile, or plan, was for Voyager 1 to be launched several weeks after Voyager 2 and to overtake the former long before the two reached Jupiter. Each probe was assigned a unique trajectory in order to accomplish specific science objectives and to obtain maximum results during the encounters with Jupiter and Saturn. Also, each spacecraft was programmed to orient its on-board navigation systems approximately every 50 million miles of flight by spinning slowly to calibrate its instruments and take optical observations in various directions.[31]

Voyager 1 was expected to begin its first encounter experiences when about 50 million miles from Jupiter, in December 1978. By February 1979 the ship was expected to be about 18 million miles from Jupiter, and close encounter would begin in March 1979 with a study of Amalthea, the innermost of Jupiter's then-known satellites, at a distance of some 258,000 miles. Its closest approach to Jupiter's cloud tops would be about 173,000 miles. Voyager 1 was also to investigate Io from a very close distance of 13,700 miles, Europa at 455,000 miles, Ganymede from an altitude of 71,500 miles, and Callisto at a distance of 77,000 miles. Voyager 2, lagging somewhat behind, would pass Jupiter at a much more distant 400,000 miles and would be more distant from each of the satellites except Ganymede, which it would pass at approximately 34,000 miles. The idea behind this difference was to protect Voyager 2 from radiation hazards and orbital debris that might be experienced in the closer approaches. If Voyager 1 failed in its scientific surveys, Voyager 2 could be redirected to accomplish those missions; otherwise, Voyager 2 was being safeguarded to better assure a successful Saturn mission and then to begin an extended mission to the outermost planets.[32]

Jupiter's gravity was to "slingshot" the Voyager spacecraft around the planet toward Saturn. One year after leaving Jupiter, in late August 1980, Voyager 1 was to begin its first studies of Saturn, and here science really was entering the realm of the unknown. The closest approach to Saturn would be about 85,800 miles, but primary targets were the planet's prominent rings and its giant moon, Titan. Depending on the results of Voyager 1's Saturn encounter, Voyager 2 was given two major options. The first was to preserve it as a "healthy spacecraft" by foregoing a close encounter with Titan, but to encounter at safe distances the satellites Rhea, Tethys, Mimas, and Enceladus, and Dione. Another option, given unsatisfactory results by Voyager 1, was to commit Voyager 2 to a close encounter with Titan and in effect to duplicate the Voyager 1 Saturn trajectory.[33] That option would likely mean losing the opportunity to go on to Uranus or Neptune.

The decision to try for Uranus, and in effect for the Grand Tour of the outer planets, would not be made until 1981, four years after launch. Then, *if* it happened, the Voyager spacecraft would reach the seventh planet from the Sun in January 1986. By all measures, the Voyager Jupiter-Saturn mission plan was a carefully constructed, detailed, critically designed and remarkable flight plan. A critical part of the mission design, as had been true in the engineering design work, was that one or both Voyagers would not only successfully complete their Jupiter-Saturn missions, but would continue on to the outer planets.[34] But, to be sure, back on Earth in August 1977, the focus was on launch and the initial flight to Jupiter.

The watch passed to Mission Director Dick Laeser and Mission Operations, who were responsible for the spacecraft, ground and science operations, and preliminary science data analysis. Although launch took place in Florida, flight operations were managed at JPL in Pasadena. Even in the early launch phases, while the operations team was still in Florida, telemetry data came in through JPL's Deep Space Network, and guidance and control directives went to the spacecraft from JPL.[35] As previously discussed, JPL's Deep Space Network was a critical ingredient in the success of the Voyager missions; it was an essential part of the unique experience and expertise that JPL brought to the program of planetary exploration.

The science program and data analysis were under the auspices of the project scientist, Ed Stone, and the Science Steering Group, also headed by Stone. The project scientist, a multimission systems office to handle the tracking data and correlate the two Voyager flights, JPL's public affairs office representative, and Kohlhase, the mission planning manager, reported to the mission director. The line of responsibility passed from Mission Operations to the Sequence Design and Integration Directorate under F. M. Sturms Jr., the Science Data Analysis Directorate headed by James E. Long, Space Flight Operations Directorate under M. Devirian, and the Science Data Analysis Directorate under Michael J. Sander. The various directorates included representatives from most of JPL's technical divisions.[36] Thus, flight operations, as had the Voyager engineering design and development efforts, effectively drew upon the total resources of the Jet Propulsion Laboratory.

On August 20, 1977, following years of effort that included program starts, no-starts, and redefinitions, Voyager 2 took flight, a Titan-Centaur launch vehicle lifting the spacecraft from its launch pad at Cape Canaveral and setting it upon its course to the giant planets. Within minutes of launch, telemetry data began coming into the Mission Control Center at JPL indicating a fault in the AACS. The spacecraft automatically shifted to its alternative AACS processor while Mission Operations frantically endeavored to find out what had happened and whether there had been any change in the navigational memory banks. Later, a boom extending the science package beyond the spacecraft was thought to have failed to deploy as scheduled. The problems seemed to be the result of "a willful computer that would not listen to its masters on the ground." More accurately, some of the reported symptoms were inaccurately transmitted or received, others were fixed by the automatic corrections systems aboard the spacecraft or by signals from ground control, and some simply "went away." Part of the general problem, one that persisted for some time, had to do with the spacecraft's creators understanding just what they had created.[37]

The second act of the Voyager drama began on September 5. After some initial launch-day complications, Voyager 1 rose from its launch pad and raced through the heavens to meet and pass its twin.

Although technically the two spacecraft were solitary, robotic machines, effectively they were being flown by their creators, by the hundreds of en-

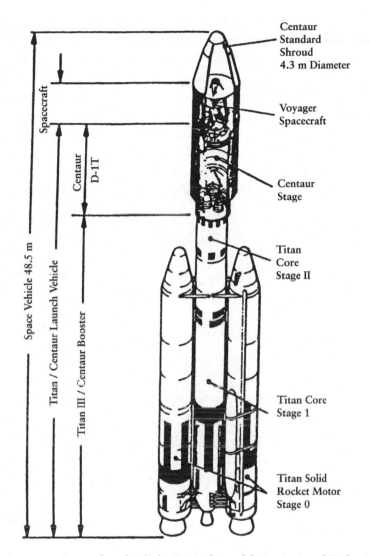

Titan IIIE–Centaur–Voyager launch vehicle. (*Proceedings of the Institution of Mechanical Engineers* 194, no. 28 [1980]: 214)

gineers, scientists, technicians, contractors and administrators at JPL and throughout the nation. Specifically, the Voyagers sought data and information about the planets Jupiter and Saturn and their space environments; and possibly one vehicle would fly on to Uranus, Neptune, or beyond. Wherever they journeyed, the Voyagers were seeking information that could shed new light on the origin and early history of the Solar System and of planet Earth, for Voyager was first and foremost a scientific mission—and it was much more.

Voyager's instrumentation, design, and hardware elicited the admiration of the engineer. Scientists shared with Voyager a sense of being on the frontiers of knowledge, and people around the world shared a sense of participation and wonderment. The Voyager missions "touched a universal, restless component of the human spirit," the longing to explore. The general attitude was that science-wise or otherwise, the planned visits to the outer planets would be "exciting and worthwhile adventures."[38]

7

VOYAGER SCIENCE

Creative engineering and innovative science characterized Voyager's exploration of the unknown. Following launch, efforts and interests shifted from engineering and design to operations and science; nevertheless, engineering remained a critical component of the Voyager mission as scientists and engineers wrestled with the unforeseen and unexpected. Project scientist Ed Stone, the interface between scientists and engineers during the spacecraft design and construction phase, now became the ombudsman among the engineers, Mission Operations, and science investigators. Stone, who became director of the Jet Propulsion Laboratory in 1991 while continuing to hold the post of Voyager project scientist, never dreamed in 1972, when he joined the team, that Voyager science would be a major part of his life's work for next three decades.[1]

His mentor, Rochus E. Vogt, had gotten Stone involved in science planning for the Grand Tour program in 1970. When Grand Tour entered into its demise, Bud Schurmeier asked him to take on the MJS '77 project scientist job. Caltech gave him a one-third release from teaching, and Stone soon found himself holding down two full-time jobs, one teaching and one with Voyager science.[2]

Voyager science proceeded from fundamentally different criteria than Voyager engineering. On the one hand, engineering had to do with building spacecraft and designing missions using very specific tools, materials, and technology. On the other hand, Voyager science was much more subjective because it involved doing something that had never been done before. The engineering mission was to build and fly a spacecraft from Earth to "giant Jupiter and ringed Saturn, several moons of both planets and possibly Uranus." In its broadest terms, the science mission was "to return information that could shed new light on the origin and early history of the solar system and our own planet Earth."[3]

But just what data could be obtained from flyby visits to those distant bodies that would address fundamental questions about the origin and nature of the Solar System and add to our knowledge of Earth? In 1977 the generally accepted theory of the origin of the Solar System was that it began with the collapse of a cloud of gas and dust to form the Sun, and remainders from that great event coalesced to form the planets, their satellites, asteroids, comets, and meteors. Astronomers and other scientists of the time loosely classified the planets as terrestrial, those resembling Earth, and the outer "gas giants," which had greater masses and lower densities. Scientists assigned an age of 4.6 billion years to the Solar System and agreed that during that time planets could have evolved. Knowledge of the composition of the planets and their satellites, as well as their temperatures, densities, magnetic fields, and infrared emissions, could contribute to a better understanding of the inception and evolution of Earth and the Solar System.[4]

But it was not so simple. Voyager science was still problematical, for how does one prepare the unknown and unforeseen? Were the science experiments and instruments selected for Voyager those really needed? How sensitive does an instrument need to be? How close does an experiment need to be to its target to obtain reliable data? How much observation time does a principal investigator (PI) need during a flyby? How long an exposure does an imaging device require? What spacecraft attitude and position are most desirable for an experiment, and how do those requirements affect other experiments and the total mission? Scientists had expectations and anticipations of what they would find at Jupiter, Saturn, and beyond, but little reliable data. Thus, Voyager scientists, Stone noted, had a lot of "desirements," but no real knowledge of their needs, because no one had been there or done that before.[5]

A rule of thumb for a principal investigator at work on a space science experiment is that "more is always better," but more is not always possible on a spacecraft, nor is it necessarily desirable. Moreover, each scientific payload aboard Voyager affected other experiments; the scientific experiments affected the design of the spacecraft, and the desired science objectives determined the spacecraft's trajectory and flight operations. The job of the Voyager project scientist was to serve as the liaison between the engineering community and the mission operations on the one hand (who wanted to be sure that the mission was possible) and the scientist as a principal investigator and scientists collectively as part of the broader science community on the other hand (who wanted to complete perceived science objectives without having full knowledge of what the real objectives or the results might be).[6]

As Bruce Murray had observed years earlier, the potential scientific discoveries in space were varied and unpredictable. The role of the project scientist was to help the project manager and the mission director make the right operational decisions. Given the many unknowns, that required numerous judgment calls. The key, Stone said, was for the project scientist to understand as well as possible what the various instruments and investigations were about. He had to know the project from the engineering and the science sides well enough to help the principal investigator understand that constraints were real and not contrived, and then help the mission managers and engineers maximize the science potential of the mission.[7] What could be learned about the outer planets of the Solar System?

What knowledge *did* astronomers and scientists have of the Solar System before the flight of the Voyagers, and how could Voyager affect that knowledge? Science was what the Voyager missions were about, and Voyager science sought to address fundamental questions about the origin and nature of the Solar System in order to obtain a better knowledge and understanding of Earth. Previous NASA missions to Mars, Venus, Mercury, and the Moon had expanded our body of knowledge and whetted our appetites for more.[8] More specifically, Voyager science had to do with knowledge about Jupiter, Saturn, and the interplanetary space between Earth and those bodies. Possibly, Voyager science might have to do with Uranus and the space beyond. But the focus of Voyager science, as had been the focus of the spacecraft design, was on Jupiter and Saturn.

As mentioned earlier, in 1973 and 1974 Pioneers 10 and 11 had sped past Jupiter and had revealed a complex atmosphere and much higher densities of energetic electrons than were expected. Jupiter has 318 times the mass of Earth and Saturn 95 times that of our planet. Jupiter itself contains more matter than all of the other planets of the Solar System combined, and with its retinue of thirteen (then-known) moons, was the center of its own miniature solar system. Saturn also boasted a coterie of small satellites, and one giant moon, Titan, whose atmosphere might be described as "massive as that of Earth's." Saturn's unique feature was the spectacular ring system, which was thought to be composed mainly of tiny pieces of water ice. Much less was known about the more distant planets other than that they had low densities compared to that of Earth.[9] The truth, as it turned out, was that we knew much less about Jupiter, Saturn, and the outer planets than we thought.

Voyager science dealt with obtaining data and then evaluating and understanding it, and to accomplish those tasks NASA drew upon a broad spectrum of the American scientific community. Space science, like space flight, was a large-scale, multifaceted enterprise, all the elements of which were channeled through JPL's Voyager Project Office and the Voyager project scientist.

The Voyager science organization began with oversight by NASA's Office of Space Science, headed, at the time of the Voyager launch in 1977, by Noel W. Hinners, who reported to Administrator James C. Fletcher. Hinners's two deputies, Anthony J. Calio and S. Ichtiaque Rasool, provided administrative assistance, and A. Thomas Young directed NASA's Lunar and Planetary Programs. Rodney Mills was Voyager program manager, and Arthur Reetz Jr., his deputy; John Casani, JPL's Voyager project manager, reported to Mills. Milton A. Mitz, Ed Stone's counterpart at NASA Headquarters, was Voyager program scientist and was most directly involved in mission operations following launch. As the Voyager flight support manager, Earl W. Glahn provided liaison among the launch, operations, and science groups.[10] In addition, other NASA administrative branches and centers were actively involved in Voyager launch, science, and flight operations (see the appendix).

Voyager management involved a far-flung and diverse association of individuals, agencies, and talents, particularly so when it came to the science experiments. Thus, NASA's Office of Tracking and Data Acquisition, under Associate Administrator Gerald M. Truszynski, derived its flight data for the most part from JPL's Deep Space Network. NASA's Office of Space Flight, administered by John F. Yardley, had overall responsibility for the Voyager launch vehicles—the Titan-Centaur, managed by Lewis Research Center. Kennedy Space Center managed Voyager launches. The federal government's Energy Research and Development Administration (through its director, Douglas C. Bauer, and assistant director for space applications) also participated in Voyager launch and mission operations. And there were numerous other agencies, educational institutions, and American businesses associated with the Voyager missions through the science and/or engineering elements of program.[11]

Hands-on management of Voyager science was the responsibility of JPL's Project Science Office headed by Ed Stone. James Long, who had headed JPL's Office of Plans and Programs, and had been one of the architects of the Mariner Jupiter-Saturn program, worked with Stone as the project science manager. Long also headed the Science Data Analysis Directorate, which gathered and distributed incoming data from the Imaging and Radio Science Teams for analysis and evaluation. The Science Office was staffed by a representative from each of the science experiments, who worked in the office variously on part-time, half-time, or full-time bases and reported to the science manager. These representatives provided continuing interfaces with the eleven science teams whose members were, in turn, located at various places in the United States and abroad and provided Stone with the hands-on knowledge and experience to determine how things actually worked on the experiment level. The Science Office provided whatever support might be required by the science teams and investigators, gave direction and oversight for the preparation and the application of the science experiments, and resolved conflicting science requirements, which in the operations mode would primarily involve radio and power needs, trajectories, on-board sequencing and memory usage, and articulation and control systems. In the final analysis, the project scientist made the scientific judgments for the project, served as Voyager's scientific spokesman, managed the project's science budgets and administrative affairs, and generally directed the science activities of the Voyager missions.[12]

The Science Steering Group, chaired by Stone, included the team leaders (TLs) or principal investigators of each of the Voyager experiments, and it functioned as the Voyager science executive committee. In the sense that each member of that committee generally championed the cause of their own experiment, Voyager science tended to be an individual and not a community issue. However, a sense of community and common cause generally prevailed over more narrow interests, thanks in good measure to Stone's management expertise: his skill, knowledge, judgment, and ability to communicate with all science interests. Because his office had a representative on each science team, Stone was better able to understand how the science really worked and to determine the real needs and issues relating to the science experiments and with that information he could better accommodate the needs of the science investigators while also sustaining the missions' basic engineering and operations requirements.[13] Indeed, although the science involved in Voyager professed to be "hard" and objective, managing that science was an art.

For the most part, as previously noted, Mariner Jupiter-Saturn science experiments became Voyager science experiments, but there were some changes and modifications along the way. The most prominent change followed the revelations from Pioneer 10 and 11 that Jupiter's environment contained unexpectedly high intensity levels of energetic electron radiation. That finding forced a complete reevaluation of all Voyager experiments and the mission trajectory, as well as a redesign of the spacecraft to increase radiation hardening. In fact, Stone said, as it was originally designed the Voyager spacecraft could not have survived its Jupiter encounter. Because the Jupiter radiation environment was so much harsher than expected, the Science Office decided to refine an original plasma study headed by Herbert Bridge at the Massachusetts Institute of Technology (MIT) and add two independent but related experiments. Bridge's team was to focus on interstellar space, and a previously scheduled interplanetary/interstellar particulate matter experiment (to be conducted by Robert K. Soberman of Drexel University in collaboration with General

Electric) was dropped. Stone added a new Plasma Wave Team headed by Frederick L. Scarf, a scientist with TRW Systems and Donald A. Gurnett at the University of Iowa. This team was to target planetary environments and that of Jupiter in particular.[14] So, at launch time, with the exception of the addition of the plasma wave experiment and the deletion of a dust detector, Voyager science had changed little since the initial selection of experiments.

The Voyager experiments were all team projects, with team members representing a broad spectrum of agencies, institutions, and firms, and were selected because they were considered by NASA and the American scientific community the best studies that could be included. The selection process began with the issue of what was then called an announcement of opportunity (often followed with a request for proposals) by the NASA Office of Space Science. That announcement was itself the product of a science working group commissioned by NASA to identify issues and scientific studies that might produce the most meaningful and productive returns from a flyby study of Jupiter and Saturn. The announcement of opportunity identified areas of study recommended by the working group, and invited scientists to submit a proposal to be a team member, or to build their own instrument to study areas that were identified as being of primary interest to the mission.[15]

Once proposals were submitted, they were distributed by NASA to special peer review committees comprised of independent scientists who could pass on the validity of a proposal, assess its feasibility, and assign some priority to the proposed investigation. By the time it came to choose experiments for the Voyager missions, NASA's process for doing so had benefited from over a decade of experience in such matters. Most of the proposers were familiar with that process, a situation that helped greatly in eliminating charges of cronyism, bias toward a specific field of study, and the like. Then, too, the decision to include teams of scientists, rather than lone researchers, meant that many more individuals could be satisfied. Of course, the announcement of opportunity alerted interested researchers to the types of science NASA considered important, though this was in many ways "fuzzy." After all, the Voyagers were heading into mostly unknown territory, where nobody knew just what to expect.

More rigorous restraints were imposed by practical engineering considerations; an instrument weighing in at one hundred pounds, or one requiring a kilowatt of electrical power, could not be flown. There also were many questions concerning how particular experiments would affect others or the operation of the spacecraft itself. An instrument to measure magnetic fields, for example, must be positioned at some distance from the main body of the spacecraft and its electrical currents and associated magnetic fields. The answer was to mount it on a long boom, but how would that affect the probe's ability to change its orientation or trajectory?[16]

After studying the committees' recommendations, selection of the actual experiments to be flown was then made by the Office of Space Science, which subsequently approved any recommended modifications suggested by Stone's Voyager Project Science Office.[17] Thus, at launch, there were eleven Voyager science investigations, each supported by a team of scientists and each team chaired by a principal investigator or team leader.

Bradford A. Smith, for example, who headed the Imaging Team, was completing work on his doctorate degree in planetary astronomy at New Mexico State University when the Voyager announcement of opportunity was issued. Smith had taken the "back roads" into the field of astronomy. He remembered that as a child of eight or nine living in Cambridge, Massachusetts, his grandmother gave him a book on astronomy, and it captured his fancy. He got in the habit of biking to Harvard College Observatory to see the telescopes and astronomers at work. But he also remembered that one of the most prominent astronomers there, Fred Whipple, lived out of the back of his Model A Ford and wore patches on the sleeves of his coat before patches became fashionable. Smith decided that a bachelor's degree from Northeastern University in chemical engineering offered far greater opportunities than astronomy, but he had only buried, not obliterated, his appetite for astronomy.[18]

The military called about that time, and Smith was assigned to the Army Map Service and from there he was sent to a special detail with the Office of Ordnance Research to study the processes of locating specific points on Earth by using artificial satellites as reference points. In that work he became closely associated with Clyde W. Tombaugh, a former amateur astronomer and discoverer of the planet Pluto. His friendship with Tombaugh and the work related to celestial navigation rekindled his

Left to right: Charles Kohlhase, Clyde Tombaugh, and space artist Don Davis, November 1980. (Courtesy of Charles Kohlhase)

interest in astronomy, and Smith decided that whether astronomy "paid or not," that was what he really wanted to do. So, after he left military service he began the lengthy process of work on a doctorate degree in astronomy, which he completed at the age of forty-three.[19]

Near the close of his studies, and just before accepting a position at the University of Arizona as the observational astronomer (replacing Gerard P. Kuiper, who had recently died), Smith submitted a proposal to serve as a member of the Voyager Imaging Team and checked the square indicating that he would be willing to serve as the team leader. Sometime later Milton Mitz called to tell him that not only had he been selected for the Voyager Imaging Team, but he was to be the team leader if he would accept the job. He did. Smith was then informed that the instrument to be used for the imaging work was one already acquired by NASA. "You've got an instrument and you're stuck with it, and you are not going to change it, and that's that!" Mitz informed him. "So of course," Smith confided, "at the very first meeting we had we went about changing our instrument because it was an instrument designed to study the planets, and no one had thought that much about the satellites. We realized that the satellites were going to be extremely important and we needed an instrument with a focal length that could look at satellites."[20] As it turned out, contrary to the expectations of many in the science community, the imaging studies—in particular those of satellites—proved to be among Voyager's most significant accomplishments.

Indeed, imaging of planetary bodies had advanced considerably since the not-too-distant days when Lowell Observatory churned out "image after image" of Jupiter and Saturn, "pretty pictures that no one ever did any science with." Photography held little esteem among astronomers because the photographs were only "pictures" with little or no quantitative scientific value. That attitude, Smith recalled, prevailed well into the space age, but as the technology improved and digital images became available, the science community began to appreciate that an image was actually an array of photometric points. Before long, imaging, already popular with the media and the public, became scientifically acceptable as a quantitative study.[21]

Voyager, Smith thought, was very different from other NASA missions. It was "a group of people really interested in science." He recalled that during meetings of the Science Steering Group, on many occasions one principal investigator voluntarily gave up allotted observing time to another because of the realization that that work at that time was more important than what his own team would be doing. The usual backbiting and unpleasant interactions were diminished. "Voyager was a very big team," he said.[22]

Garry E. Hunt, who worked with Smith on the Imaging Team and was the sole scientist selected from the United Kingdom by NASA, echoed that evaluation. "Voyager," he said, "is a very special collection of people" more so than anything else; it became "a community, a family, an involvement" for a lifetime.[23]

Voyager was a scientific community in itself, and it reflected fundamental changes that had occurred in science in general and in astronomy in particular over the past quarter-century. The "Edison-Lowell era" of independent science and discovery, of the lone-observer hunkered down over the microscope or under a telescope, had given way to team science, to "big science" and "big technology" with costly equipment, huge investments in personnel, large budgets, bureaucracies, and the anticipation of large payoffs and products. Historian Andrew Butrica styled Voyager "archetypical big science" because of its "scientific, budgetary, and technological immensity," as well as a prototype of the developing "NASA-industrial-academic complex."[24] In particular, Voyager engaged individual scientists as members of science teams, which collaborated with other teams and NASA centers and with industry, through the NASA/JPL management nexus.

The essence of Voyager science were the eleven teams headed by a principal investigator or team leader. At launch, those teams included the following individuals (over the course of the next several decades, the lists changed, as Voyager was a multigeneration science project):

IMAGING SCIENCE: TL, BRADFORD SMITH, UNIVERSITY OF ARIZONA

The imaging science equipment included two TV cameras with 1500-mm f/8.5 and 200-mm f/3 optics, with multiple filters and a wide-angle field, all mounted on an independently maneuverable scan platform.

Members	*Affiliation*
Geoffrey A. Briggs	Jet Propulsion Laboratory
Alan F. Cook	Smithsonian Institution
George E. Danielson Jr.	Jet Propulsion Laboratory
Merton Davies	Rand Corporation
Garry E. Hunt	Meteorological Office, U.K.
Harold Masursky	U.S. Geological Survey

Tobias Owen	State University of New York
Carl Sagan	Cornell University
Laurence Soderblom	U.S. Geological Survey
Vernon E. Suomi	University of Wisconsin

INFRARED RADIOMETRY AND SPECTROSCOPY: PI, RUDOLF HANEL, GODDARD SPACE FLIGHT CENTER

The spectrometer-radiometer (IRIS), also mounted on the scan platform, could measure temperatures and molecular gas atmospheric compositions, of both planets and their satellites.

Members	*Affiliation*
Barney J. Conrath	Goddard Space Flight Center
Peter Gierasch	Cornell University
Virgil Kunde	Goddard Space Flight Center
Paul D. Lowman	Goddard Space Flight Center
William Maguire	Goddard Space Flight Center
John Pearl	Goddard Space Flight Center
Joseph Pirraglia	Goddard Space Flight Center
Robert Samuelson	Goddard Space Flight Center
Cyril Ponnamperuma	University of Maryland
Daniel Gautier	Meudon, France

ULTRAVIOLET SPECTROSCOPY: PI, A. LYLE BROADFOOT, KITT PEAK NATIONAL OBSERVATORY

The ultraviolet spectrometer, mounted on the scan platform and weighing slightly less than ten pounds, enabled scientists to measure ionic, atomic, and small-molecular-weight gas abundances, analyze the upper atmospheres of Jupiter and Saturn, measure the absorption of the Sun's ultraviolet radiation in the upper atmospheres of the planets, and determine the distribution and ratio of hydrogen and helium in interstellar space.

Members	Affiliation
Jean-L. Bertaux	Service d'Aeronomie du Centre National de la Recherche Scientifique (CNRS), France
Jacques Blamont	Service d'Aeronomie du CNRS
Thomas M. Donahue	University of Michigan
Richard M. Goody	Harvard University
Anthony Dalgarno	Harvard College Observatory
Michael B. McElroy	Harvard University
James C. McConnell	York University, Canada
H. Warren Moos	Johns Hopkins University
Michael J. S. Belton	Kitt Peak National Observatory
Darrell F. Strobel	Naval Research Laboratory

PHOTOPOLARIMETRY: PI, CHARLES F. LILLIE, UNIVERSITY OF COLORADO

The photopolarimeter was a telescope with variable aperture, filters, polarization analyzers, and a photomultiplier tube detector, mounted on the scan platform, that analyzed the way in which planets and satellites reflect light, thus providing information on the surface structure and composition of a body. The instrument would study aerosol particles in the atmospheres, textures and compositions of satellites, shape and structure of Saturn's rings, and search for interplanetary and interstellar particles.

Members	Affiliation
Charles W. Hord	University of Colorado
Douglas L. Coffeen	Goddard Institute for Space Studies, New York
James E. Hansen	Goddard Institute for Space Studies, New York
Kevin Pang	Science Applications, Inc.

PLASMA SCIENCE: PI, HERBERT S. BRIDGE, MIT

Plasma are clouds of ionized gases moving through interplanetary space and produced by the Sun and other stars. The plasma science investigation employed two Faraday-cup plasma detectors to study the solar wind, planetary magnetospheres, interstellar ions, and the satellites of Jupiter and Saturn.

Members	Affiliation
John W. Belcher	MIT
John H. Binsack	MIT
Alan J. Lazarus	MIT
Stan Olbert	MIT
Vytennis M. Vasyliunas	Max Planck Institute, West Germany
Leonard F. Burlaga	Goddard Space Flight Center
Richard E. Hartle	Goddard Space Flight Center
Kieth W. Ogilvie	Goddard Space Flight Center
George L. Siscoe	University of California–Los Angeles
Alan J. Hundhausen	High Altitude Observatory

LOW ENERGY CHARGED PARTICLES: PI, TOM M. KRIMIGIS, JOHNS HOPKINS UNIVERSITY

The LECP experiment employed a low-energy magnetospheric particle analyzer and a low-energy particle telescope to examine Jupiter and Saturn's magnetosphere and charged particles associated with their satellites and Saturn's rings; to measure components of the solar wind; and to study galactic cosmic rays from outside, and solar particles within, the Solar System.

Members	Affiliation
Thomas P. Armstrong	University of Kansas
W. Ian Axford	Max Planck Institute
Carl O. Bostrom	Johns Hopkins University
Charles Y. Fan	University of Arizona
George Gloeckler	University of Maryland
Louis J. Lanzerotti	Bell Telephone Laboratories

COSMIC RAY: PI, ROCHUS E. VOGT, CALIFORNIA INSTITUTE OF TECHNOLOGY

The cosmic ray investigation used a low-energy solid-state detector telescope, a high-energy telescope, and an electron telescope system to measure the energy spectrum of electrons and cosmic-ray nuclei, to study radiation in the environment of Jupiter and Saturn, and to determine the intensity of energetic particles as a function of distance from the Sun.

Members	Affiliation
J. Randy Jokipii	University of Arizona
Edward C. Stone	Caltech
Frank B. McDonald	Goddard Space Flight Center
B. J. Teegarden	Goddard Space Flight Center
James H. Trainor	Goddard Space Flight Center
William R. Webber	University of New Hampshire

MAGNETIC FIELDS: PI, NORMAN F. NESS, GODDARD SPACE FLIGHT CENTER

The four magnetometers aboard Voyager were to determine the magnetic field and magnetospheric structure at Jupiter and Saturn, the interaction of magnetic fields and satellites orbiting within those fields, as well as to study the interplanetary-interstellar magnetic field.

Members	Affiliation
Mario H. Acuna	Goddard Space Flight Center
Kenneth W. Behannon	Goddard Space Flight Center
Leonard F. Burlaga	Goddard Space Flight Center
Ronald P. Lepping	Goddard Space Flight Center
Fritz M. Neubauer	Technische Universitat, West Germany

PLANETARY RADIO ASTRONOMY:
PI, JAMES W. WARWICK,
UNIVERSITY OF COLORADO

The Planetary Radio Astronomy investigation consisted of a stepped frequency radio receiver and two monopole antennas (33 feet long) to detect and study a variety of radio signals emitted by Jupiter and Saturn.

Members	*Affiliation*
Joseph K. Alexander	Goddard Space Flight Center
Andre Boischot	Observatoire de Paris, France
Walter E. Brown	Jet Propulsion Laboratory
Thomas D. Carr	University of Florida
Samuel Gulkis	Jet Propulsion Laboratory
Fred T. Haddock	University of Michigan
Chris C. Harvey	Observatoire de Paris, France

PLASMA WAVE: PI, FREDERICK L. SCARF,
TRW DEFENSE AND SPACE SYSTEMS

The three-pound plasma wave instrument measured thermal plasma density, wave-particle interactions, and the interactions of satellites of Jupiter and Saturn with their planets' magnetospheres.

Members	*Affiliation*
Donald A. Gurnett	University of Iowa

RADIO SCIENCE: TL, VON R. ESHELMAN,
STANFORD UNIVERSITY

This was the only science investigation that did not have its own dedicated science instrument. The spacecraft's telecommunications system was used to measure the way radio signals faded and returned when the spacecraft disappeared behind a planet or satellite and then reappeared. Such behavior helps to establish the properties of planetary and satellite atmospheres and ionospheres, and measures the spacecraft's trajectory and Saturn's ring mass, distribution, and structure.

Members	Affiliation
John D. Anderson	Jet Propulsion Laboratory
Thomas A. Croft	Stanford Research Institute
Gunnar Fjeldbo	Jet Propulsion Laboratory
George S. Levy	Jet Propulsion Laboratory
G. Leonard Tyler	Stanford University
Gordon E. Wood	Jet Propulsion Laboratory

Managing those teams to create a consolidated community effort, and to maximize the potential of each scientific investigation, involved both art and expertise. Stone's Voyager Science Office worked closely with Kohlhase's Mission Planning Office to assure that the optimum mission profile was being maintained, and that corrections and adjustments to flight trajectories could be made to enhance the science requirements. The actual integration of science and flight engineering requirements was done by the Sequence Design and Integration Directorate, under Francis M. Sturms Jr. His office plotted the preflight operating schedule, which included science operating time allotments, communications access, and maneuver sequences as required by the science investigators, after integrating those requests with the spacecraft engineering, navigation, and ground operations requirements generated by the Space Flight Operations Directorate headed by Michael Deverian as well as the general flight constraints established by the Mission Planning Office. Following launch, the flight schedule was constantly adjusted to accommodate new data and requirements.[25]

The Science Integration Team and its chief, Arthur Lonne Lane, was central to the Voyager science mission. The group included representatives from each of the science investigations, members of the Science Observation Analysis Team (that is, those who appraised the integrity and value of the data), the sequence mission engineer, science sequence planners and science sequence coordinators. The team established the overall mission time line and resolved conflicts between requirements of competing scientific observations. It cleared its recommendations with the Science Office and the Science Steering Group and negotiated with the Mission Planning Office for variances in the mission. During the mission the Science Integration Team processed requests from the investigating teams and transmitted those requests, if it concurred, to the Space Flight Operations directorate for changes to the flight sequences. The Science Integration Team was, in effect, the "science advocate" for the mission and was the first management group for establishing the community interests of the Voyager science and representing those interests to the Science Office and mission operations.[26]

Another critical management team was the Science Data Analysis Directorate, headed by James Long, which was responsible for analyzing and coordinating incoming data generated by the science experiments. About one year after launch, Long became the Science Office manager under Ed Stone; routine business and the science returns were channeled through the Voyager Science Office. Stone recalled that during Pioneer 10's Jupiter encounter in December 1973, he first became aware of the tremendous public demand for science information and for the necessity of carefully appraising "instant science" before making a public release. During that encounter the mission science representatives met each morning to assess the data and then immediately adjourned for a public press conference. It was the first time in his science career, Stone said, that he had ever seen a time and place where the media wanted to know what you were discovering as you did it. Usually, NASA grants exclusive rights to data obtained from a space mission to the scientists directly involved for a specific period, typically a year or so. But lunar and planetary exploration proved different from the start. These situations created an unusual opportunity for a scientific endeavor to communicate the process of discovery and to engage the public in some of the excitement that scientists shared, but it also allowed the possibility of conveying misinformation and creating confusion.[27] Stone was determined to establish careful procedural guidelines and structures that would distill, analyze, and evaluate data prior to public release.

Stone's problem was to explain the science and new discoveries to the media in such a way that they could write about it, for the information could not be too technical, but had to be descriptive enough to engage the interests of the general public. The day began with an 8:00 A.M. meeting at which the principal investigators or their representatives would report on the status of their investigations, and a press conference followed at 10:00. Stone imposed a rule that nothing discussed in the earlier status session could be presented at the press conference because there had not been adequate time to digest the information. He then convened a meeting of the entire science team at 1:00 P.M., often including as many as one hundred people from various disciplines, to conduct a form of peer review of the incoming data. At that session scientists reviewed and discussed what they were thinking and how they evaluated their own data and that reported by their colleagues. The meeting closed with decisions on which things were ready to present to the media at the next day's press conference. By that time, the information released to the media, though current, would have been as focused and as accurate as possible. It was Stone said, a daily process—"everyday!"—during the encounter episodes.[28]

Stone concluded that the benefits of this process went far beyond an effective media presentation, for it enabled the various instrument teams to talk to each other about their science in a peer review process. The net result was that Voyager science was a true team effort rather than the product of eleven individuals or disparate groups who got up and surprised their colleagues and the world with their discoveries, as not infrequently occurs in the competitive world of scientific research. Instead, Voyager team members shared parts of a wonderful and remarkable experience.[29] It was the job of the Voyager project scientist to communicate and share this experience effectively with the media and the public.

Voyager became an adventure in creative engineering, innovative science, and astronomical surprise; what scientists thought existed, and predicted, often failed to be, for nature proved to be much more diverse than human ideas. In space, nature seemed to be able to apply the same physical laws known to exist on Earth to produce things remarkably unearthly. Voyager was (and still is) a learning experience.

What did scientists expect to learn from the Voyager mission? According to pre-launch press information, the Voyager mission to Jupiter and Saturn was to "address fundamental questions about the origin and nature of the Solar System." Previous missions to Mars, Venus, Mercury, and the Moon had already contributed greatly to our body of knowledge; each of these "terrestrial" planets is unique. However, it was known that Jupiter and Saturn and the other outer planets differed significantly from the terrestrial variety in that they had much lower average densities and much greater masses.[30] A study of these planets, it was believed, would provide a clearer picture of the "4.6-billion-year evolution" of the planets.

More specifically, scientists and astronomers anticipated that the Jupiter encounter would help explain that planet's great mass and its mysterious "Great Red Spot" and provide hard data on its four great Galilean moons, including Io, Europa, Ganymede, and Callisto, and its smaller and more elusive satellites. The composition and particle size of Saturn's rings were unknown, and little was known about its giant satellite Titan, other than that it existed.[31] The truth was that science knew little about the planets, and it knew less the more distant those planets were from Earth. The Voyager missions were indeed voyages of discovery.

Launch of Voyager 2 on August 20, 1977.
(NASA photo 101-KSC-77PC-270.)

Right: Goldstone antenna at Camp Irwin,
California, part of JPL's Deep Space Network.
(NASA photo 332-8661bc)

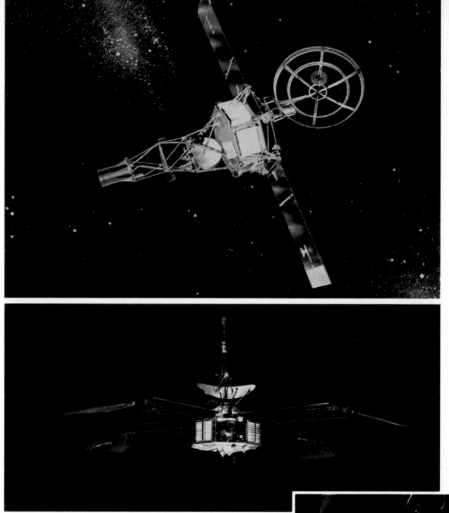

Top, above: Mariner 2, the world's first successful interplanetary spacecraft. (NASA photo Venus/P1953)

Above: Mariner 4 provided the architectural plan from which the Voyager spacecraft were designed. (NASA photo Spacecraft/P4732)

Right: Charles Kohlhase. (NASA photo P-16327A)

TITAN/CENTAUR COMPLEX 41

Voyager 2 on the launch pad,
August 1977. (NASA photo PIA-
01480)

Right: Voyager spacecraft in the
deployed position. (NASA photo
S-79-31146)

Voyager spacecraft encapsulated in Centaur standard shroud. (NASA photo JPL-373-7537AC)

Right: Solar winds become turbulent at the termination shock, and interstellar space is encountered beyond the heliopause. (NASA photo montage)

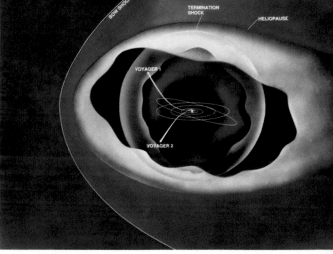

Collage of Jupiter and its four large
moons (not to scale but in relative positions). (NASA photo PIA-01481)

Right: Jupiter, from Voyager 1, March 1
and 3, 1979. (NASA photo PIA-01512)

Below. Jupiter's magnetic fields. (NASA
photo 260-522AC)

Above: A photographic replication of the Saturnian system. (NASA photo PIA-01482)

Left: Saturn, taken by Voyager 1 early in the encounter, October 1980. (NASA photo PIA-02225)

Opposite, above: Methane gas clouds on Uranus, taken by Voyager 2 in 1986. (NASA photo PIA-00143)

Opposite, middle: Artist's sketch of Voyager's 2 encounter with Uranus. (NASA photo P-23836A)

Opposite, below: Artist's conception of Voyager as it passed Saturn and continued to Uranus. (NASA photo montage PIA-02973)

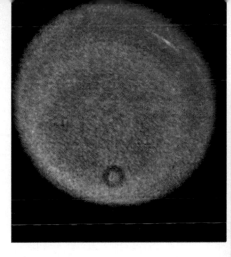

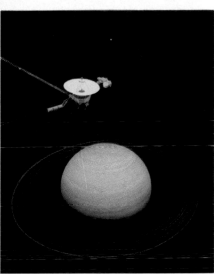

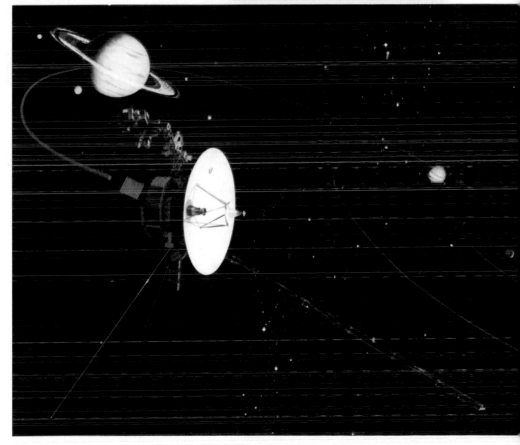

Montage of the planets and four of Jupiter's moons against a false-color Rosette Nebula (Earth's moon in foreground). (NASA photo PIA-02973)

Left: The far side of Uranus, taken by Voyager 2 on January 28, 1986. (NASA photo PIA-00143)

LAUNCH

AUGUST/SEPTEMBER 1977

One of the most critical phases pertaining to the success or failure of Voyager science was, in fact, the launch of the spacecraft from Earth. For that reason, it is important to take a close look at those few critical minutes when the performance of a complex and fickle rocket vehicle determines the fate of a mission. All the imagination, dedication, and hard work of scientists and engineers alike amounts to little if the payload ends up as a pile of junk on the floor of the Atlantic Ocean off the coast of Florida.

Missions sometimes fail suddenly and catastrophically somewhere in space, but such fiascos do not have the shattering impact of a sky full of fire raining down on Earth. Alas, there is no escaping the fact that booster rockets by their very nature contain an enormous amount of concentrated energy in their propellants and are highly temperamental beasts at best.[1]

Although NASA had abandoned the mighty Saturn V launch vehicles, the modified Titan IIIs used for the Voyager launches were powerful rockets indeed. Their genealogy began with the Titan I, a two-stage intercontinental ballistic missile designed to carry relatively heavy (by U.S. standards) thermonuclear warheads and using cryogenic propellants such as liquid oxygen. The need for a refrigeration plant next to each launch site reduced the effectiveness of such rockets as practical weapons, so the next version, the Titan II, boasted two stages using propellants that could be stored at room temperature. This latter version, with a uniform diameter of 10 feet, became the basis of the Titan III family of launch vehicles, which might or might not use a combination of a wide variety of additional upper stages and solid-fueled, "strap-on" booster stages. The Titan I and Titan II had only short careers as weapons, but the Titan III soldiered on and, in various developed forms, continues in use into the twenty-first century for spacecraft launches from the Kennedy Space Center in Florida and the Western Test Range in California. For programs such as Voyager and the Viking Mars probes, NASA used a number of these vehicles, but the bulk of them have been used to orbit a wide spectrum of secret surveillance satellites.[2]

The modified version of the Titan III used for Voyager towered 157 feet from its base on the launch pad to the tip of the bulbous aerodynamic shroud that protected the payload during the first few minutes of flight. The complete vehicle had a daunting number of separate elements, including no less than five separate stages that had to complete successfully six separate "burns," every one having to start on time, deliver the correct thrust in the desired direction, and stop on time.

The central part, or core, of the booster was a modified version of the Titan II. The two engines of the first stage delivered a combined thrust of 500,000 pounds, whereas the lone second-stage engine produced 100,000 pounds; both burning a combination of nitrogen tetroxide (N_2O_4) and a compound called Aerozine 50. These liquid propellants have the advantage of being hypergolic, that is, they burst into flame spontaneously when brought into contact and, therefore, need no ignition mechanism—thus one less system that might fail. However, that same property means that even a tiny leak somewhere in the complex plumbing has a good chance of resulting in a catastrophic explosion, and the substances themselves are not only toxic, but also corrosive.

Bolted to opposite sides of the basic "core" were a pair of powerful solid rockets containing aluminum powder, ammonium nitrate, and a binder, all mixed together and baked until the resultant "grain" has the consistency of rubber. The direction in which the elements of the "zeroeth" stage exert their thrust is controlled by the injection of liquid nitrogen tetroxide (carried in a separate tank) into their exhausts. This control mechanism contrasts with that of the core's first and second stages, whose engines can be gimbaled mechanically to alter the direction of thrust.

At launch the first-stage engines are started and, when and only when sensors indicate that they are performing properly (stable combustion, correct thrust levels, and the like), the solid boosters are ignited. This is the moment of no return, for once the solids "light off" there is no stopping them short of destruction by mishap or command of the range safety officer. As the solids' thrust begins to "tail off" after a few minutes, they are separated from the core by explosive bolts, shoved away from the rest of the vehicle by small solid-rocket motors, and fall back to the ocean. Next, the first stage shuts down, separates, and falls away as its propellants near exhaustion; the second stage then takes over.

The third stage, mounted atop the basic core, is a Centaur system whose twin engines burn liquid hydrogen and oxygen. The former is an especially difficult propellant to work with and, combined with the stage's extremely lightweight structure, resulted in a long and trouble-filled development period for Centaur. In addition, this extremely efficient yet unforgiving system has a number of novel and complicated features, such as external insulation panels that prevent the cold liquid hydrogen from boiling away before it is used, but which must be jettisoned once above the frictional heating effects of Earth's atmosphere, as well as a single insulated wall separating the liquid hydrogen and oxygen tanks. Mounted on the Centaur are the electronics needed not only for its own guidance and control, but also for those of all the lower stages.

The Centaur stage fires twice, once to put it and a payload into a low Earth orbit and again after an interval of ten to forty-five minutes of coasting to attain escape velocity from our planet. Restarts of large liquid-propellant rocket engines in "free fall" present special difficulties, and all too often lunar and planetary payloads have ended up trapped in "parking orbits" only a hundred miles or so above the ground when such restarts fail.

Even if the Centaur performed well, however, Voyager still could not attain a high enough velocity to get to Jupiter. The job required yet another propulsive stage, one providing 15,000 pounds of thrust for forty-five seconds and adding almost 4,500 miles per hour to the payload's speed. This boost was provided by a "propulsion" or "injection" module, whose main component was a solid-propellant motor derived from one developed years earlier as an upper stage for the relatively small Delta launch vehicle and which was a "stretched version" of the main retro rockets used on the Surveyor lunar landers. The module relied on the injection of liquid propellant to direct its thrust (the same method used in the zero-stage solids) and had small, separate, roll-control engines, but was directed by electronics aboard the Voyager probe, rather than by its own or those aboard the launch vehicle. Indeed, the payload and this last boost stage were treated as a single "powered spacecraft" until the empty rocket casing was jettisoned a short time after its mission was completed. This separation itself was tricky, for it had to be done in such a way that the spent booster not only would not collide with the probe proper but also had a vanishingly small chance of hitting or even passing near any planet or known satellite.[3]

The chances of all those rocket stages working as planned and the payloads functioning properly at all, let alone for years, were risky, for much less ambitious deep-space missions had failed more often than not before Voyager. Failure rates could be estimated, of course (and after the fact they could be computed exactly), but systems were always being upgraded to attain lower weight, greater engine thrust, more powerful computers, and so forth, so that few launch-vehicle and payload combinations were ever identical—and the improvements did not always work the first time they were tried. "Old-timers" at JPL remembered all too well missions in which various rocket stages blew up or did not restart, when payload aerodynamic shrouds failed to separate, or when flawed computer programs caused mistaken destruction in the cause of safety. Of course, technology had advanced since those earlier missions, but on the other hand the Voyagers and their launch vehicles were very much more complicated machines. Although the media and the general public had come to expect success as a matter of course, those who worked on the project were aware of the potential pitfalls from bitter experience.[4]

Amazingly, both Voyager launches were tremendous successes. At 10:29:45 A.M. Eastern Daylight Time on August 20, 1977, Voyager 2 lifted off at Cape Canaveral. During the early phases, telemetry indicated problems with the payload's attitude and articulation control system as well as with one of its three gyroscopes. As part of a fail-safe routine designed for just such a situation, once the spacecraft was flying free, it was commanded to search for and orient itself to the Sun only, leaving it stabilized about only two axes. The search for the guide star Canopus was delayed until mission control could verify that the correct computer codes for accomplishing the maneuver were indeed aboard the vehicle (Canopus was eventually acquired on August 24).[5]

All but one of Voyager 2's various booms and antennas, which had been folded against the body of the spacecraft during launch, deployed properly, but the science boom apparently had not done so. However, there were indications that it was very close to full deployment and, when the instruments mounted there were turned on, they worked just fine. The boom gave no further trouble.

Several times early into the flight the orientation of the spacecraft suddenly changed. There was concern that the spent propulsion module had collided with Voyager, perhaps because residual propellants in the spent solid motor casing had flared up after the stage had supposedly shut down, or that the pressure of sunlight and the solar wind might have done the same thing. However, close investigation ruled out that possibility. The cause or causes of these anomalies never were pinned down, but outgassing from the spacecraft, or reflections from stray specks of dust detected by the Sun or Canopus sensors were suspects.

Characteristically, the popular news media concentrated on Voyager 2's "failures," in the hoary tradition of "if it bleeds, it leads." Thus, although the trade publication *Aviation Week & Space Technology* properly titled its August 29 story of the launch "First Voyager on Trajectory to Jupiter," the *Baltimore Sun* for August 21 headlined "Voyager 'in Trouble' after Lift-off: Instrument Problem Develops on Flight to Jupiter, Saturn." The story quoted John Casani as saying, "The spacecraft is in trouble. There is considerable doubt about the health of the spacecraft" and "At the present time, things do not look too good." Fortunately, in spite of ominous headlines and stories to the contrary, Voyager 2 eventually "settled down" and performed well.[6]

In light of the problem in extending Voyager 2's science boom, extra springs were added to Voyager 1 to ensure proper deployment. This modi-

Assembling the flight model of Voyager 2. (NASA photo 260-180-A)

fication had to be done quickly, as only one of the Titan III launch pads at the U.S. Air Force Complex 41 was equipped to handle the Centaur upper stage. The launch window allowed only a short "turnaround" time between the two missions (about thirty days), during which damage done to ground facilities by the first launch had to be repaired, the second vehicle placed into position, and final check-outs given to the booster and its payload, but the job was done.

Voyager 1 left Earth at 8:56:01 A.M. Eastern Daylight Time on September 5, 1977. There were none of the problems that had dogged Voyager 2, and, except for a few minor "glitches" that were more puzzling than threatening, the launch went according to plan. As a bonus, the spacecraft's attitude-control system used only 3.6 pounds of hydrazine instead of the estimated 11.5 pounds during the propulsion module's burn, leaving that much more for use later in the long mission. Voyager 2 had benefited from a similar savings, which boded well for a possible extended mission to Uranus and perhaps Neptune.[7]

Once the two spacecraft had survived the hazardous boost phase; extended their antennas, booms, and scan platforms; acquired their celestial references and become stabilized about all three axes; and checked out their various systems, everyone involved in the mission breathed much easier. Relaxation was justified to a large extent for a number of reasons. For one thing, a deep-space probe in cruise phase is called upon to do very little, many systems being shut down or operating at very low power levels, and relatively few commands being sent from Earth. For another, the deep-space environment is, in general, benign. Of course there is always the chance of being hit by a small meteoroid or encountering an especially dense concentration of energetic particles from a solar storm, but there are substantial compensations. For example, a deep-space probe is not subjected to any aerodynamic, acoustic, or gravitational stresses. Then, too, there are no thermal stresses such as a near-Earth satellite encounters as it passes repeatedly in and out of our planet's shadow, not to mention no rain, snow, or humidity to corrode things. Of course there are critical periods in this phase, such as when midcourse maneuvers are made, but for most of the time, things literally "just float along."

However, there is such a thing as being too relaxed, and this happened with Voyager. The problem was due in no small part to the fact that JPL was then in the initial stages of work on the Galileo program, which was intended to send an entry capsule into Jupiter's atmosphere and take images and other observations of the planet and its satellites over an interval of several years. The story of what happened to Voyager is perhaps best told in the words of a "Caltech-JPL Performance Evaluation" made by NASA Headquarters in 1979:[8]

Voyager Program—After the launches of the spacecraft in the summer of 1977, the Voyager project suffered from a lack of strong project management at JPL. Both Voyager and Galileo were placed under a single project manager who devoted 90% of his time to Galileo. Key technical personnel who had worked on Voyager were assigned full time to Galileo. As a result, Voyager mission operations from October 1977 to the spring of 1978 were not conducted in a well-planned, disciplined manner. Planning for encounters fell behind schedule. Spacecraft anomalies were not investigated rapidly. In late December [1977], a complex spacecraft maneuver was performed with inadequate personnel present in the operations areas—the maneuver was aborted because of a software anomaly requiring management decisions.

In April 1978 a spacecraft was jeopardized when the flight team forgot to send a required weekly command.[9]

No doubt about it, management had become lethargic. This was a serious indictment, and NASA Headquarters exercised its oversight role in no uncertain terms. Washington controlled the Voyager program because, in the final analysis, it controlled the money, and soon there were substantial changes at JPL. In the words of the evaluation,

At the request of NASA, JPL presented a project reorganization plan in April. The Lab Director issued a priority letter for Voyager in April and the Voyager and Galileo projects were separated, each getting a project manager. The Voyager project was able to recover key technical personnel and to staff more fully all project positions. The Jet Propulsion Lab requested, and NASA approved an increase in the Voyager budget in June 1978 to accelerate encounter preparations. The overall project performance began to improve greatly. Detailed scheduling was implemented. Training activities were planned. Special spacecraft capability tests were scheduled to precede encounter operations. . . .

In May 1978, the Lab Director further augmented the project management staff by assigning the Assistant Lab Director for Flight Projects as Project Manager, with two capable, full-time deputy project managers. By July, the project was functioning smoothly with a noticeable improvement in morale. From July to September, mission operations were conducted well with good indications that the project would be prepared for the first Jupiter encounter in March 1979.[10]

Headquarters' evaluation of JPL's performance was concise:

> Several conclusions can be drawn from the Voyager experience in this time period. First, JPL management was slow to recognize the project management difficulties, or at least to react to them. Second, there was considerable difficulty in handling the preparation and testing of the very complex software required by Voyager. Third, JPL (and NASA) had clearly underestimated the complexity of flying a highly-automatic spacecraft with fault-correction routines. Fourth, there must be a carryover of expertise from the spacecraft development team to the mission operations team. There cannot be a complete turnover of personnel at time of launch. Fifth, a project cannot accomplish a complex task without proportionate funding, manpower, and talent.[11]

Conditions leading to the problems were admittedly complex—as was the mission. Despite Mariner, Apollo, and other experiences, NASA and JPL had not engaged in a mission requiring such detailed management that would extend over such a long time frame and over such vast distances. There were the experiences of the Pioneer spacecraft, but these were very much simpler exploring machines. In addition, there was an ongoing conflict between JPL's belief that they knew how to do the job better than anybody else and NASA Headquarter's oversight mandate. Moreover, there were decisions made and actions taken that, only in the in the light of 20-20 hindsight, never should have been taken.

Things soon improved, however, and, as the evaluation noted, "at the end of the period, JPL performance was generally good to excellent." Not surprisingly, this whole affair never made it onto the front pages of daily newspapers. Nevertheless, it was indeed an important matter, and it illustrates well one facet of the learning process that everyone involved in the mission had to go through.

Regardless of management shakeups, the Voyagers traveled on, and in some respects they assumed a life of their own, taking and relaying data about conditions in interplanetary space, calibrating their instruments, and adjusting pointing angles for Sun and Canopus sensors to allow for the changing geometry as the probes moved outward from our star. Back on Earth, rehearsals for the science and engineering teams that would be on duty for the actual events became important activities. Elaborate scripts were written for these dry runs, and everyone who was expected to participate in the actual event had to attend the practices. The time intervals involved were the same as the real ones were expected to be, however long those might extend. Some practice sessions involved the spacecraft themselves, extending the risks. About the only things that were not included in these comprehensive run-throughs were those events that no one could foresee—and, of course, such did occur. In any case, just like firefighters, combat soldiers, and emergency room nurses, the Voyager teams on Earth would be as well trained as time, talent, and hard work could make them.[12]

Despite the relatively tranquil cruise from Earth to Jupiter, there were a few serious problems. On February 21, 1978, the scan platform holding the "pointed" scientific instruments on Voyager 1 stuck in position. Eventually it was freed, but just in case it might stick again, cautious ground controllers "parked" it in the position that promised the best pictures during Jupiter encounter.

At the beginning of April 1978, Voyager 2 encountered an even more serious problem. When the spacecraft failed to receive a command from Earth for one week, its on-board fault protection algorithm assumed that the primary receiver had failed (which it had not) and switched the communications system to the backup receiver. "It was then," Ellis Miner recalled, "that we discovered that the tracking loop capacitor (TLC) in the backup receiver had shorted." The mission control then sent a command to Voyager to switch back to the primary receiver. When that occurred, the primary receiver blew (like a light bulb when it is turned on). "We then," Miner continued, "had no choice but to wait another week for the fault protection algorithm to again switch back to the backup receiver (thank goodness it was still enabled), and we learned to live with Voyager 2's 'tone deafness' from then on. Voyager 1's receiver was exercised regularly from then on to prevent its TLC from shorting."[13]

On December 15, 1977, Voyager 1 passed its sibling and assumed its proper position in the parade toward Jupiter. Of more importance, by late autumn 1978 both spacecraft had passed safely through the asteroid belt. Of course, Pioneer 10 and 11 earlier had done the same thing. The Pioneer and Voyager flights markedly changed scientists views on the amount of unseen hazardous material in the asteroid belt.

As the spacecraft neared their encounters with the Jovian system, it is interesting to look at just what of scientific interest they were expected to find. Whether one examines a JPL internal document, a NASA press kit, or an editorial in the journal *Science,* a few general trends stand out. The overall impression is that NASA was "playing its cards close to the vest" by emphasizing new information about phenomena that were already known to be present or were strongly suspected to be. For example, there would be studies of Jupiter's cloud patterns, atmospheric circulation, the Great Red Spot, composition of its upper atmosphere, possible aurora, magnetic field, and radiation belts. Interplanetary studies would also receive good billing, for they would provide useful scientific data during cruise phase even if the encounters failed.

Nowhere is there any indication of the dominant role that images would play in the exploration of these new worlds. In part that may have been because, in years gone by, "pictures" were looked down on as "not real science," but it also may have been because little was expected from them in the way of discoveries. Of course there would be spectacular images of Jupiter's colorful and complex cloud patterns, but that was to be anticipated. Although scientists assumed that the Galilean satellites would reveal surface features, nothing extraordinary was expected.

In fact, not much was expected in any sense from Jupiter's four largest satellites. They were in the same size range as the Moon and Mercury, and those bodies had turned out to have been long dead geologically, exhibiting mainly ancient lava flows peppered with impact craters. To be sure, the Galilean moons had relatively low densities, indicating that they contained substantial amounts of water ice. Io, especially, was known to be "funny": it was orangish, immersed in a glowing cloud of ionized atoms, in some way controlled the emission of sporadic radio bursts from Jupiter, and had measured infrared temperatures that never seemed to agree with one another. Still, these bodies did not seem like promising sites for exciting discoveries. From the other, much smaller moons, even less was expected.

As we shall see, things turned out very differently from what anyone anticipated. But that was not unusual, for discoveries by their very nature can *not* be predicted, despite all the pious hopes of planners and managers. Jupiter's system might well have turned out to be rather prosaic, but the only way to find out whether that was so was to go there and look. In the process the Voyagers became engines of discovery in the truest sense, for they found things that were not only new but also unimagined.[14]

Voyager missions had an enormous impact on both the planetary sciences and the world at large, but in order to assess and understand that impact, one must be familiar with the state of knowledge regarding our planetary system before the Voyager missions. Did Voyager truly revolutionize our knowledge of the outer Solar System? How did the Voyagers affect what we knew or thought we knew about the Solar System? How, following their launch in August and September 1977, did Voyager science affect the study of planetary astronomy as a science?

Knowledge of the universe beyond the Solar System grew astronomically during the first half of the twentieth century. For example, researchers discovered that the Milky Way is a huge, rotating spiral galaxy with our Sun located near the system's edge, and the faintly luminous, fuzzy-looking "white nebulae" that had puzzled observers for over a century were revealed as other galaxies, many millions of light-years distant from our own. Even more surprising was the finding that our universe is expanding and apparently had a beginning at some definite and determinable time in the past. Closer to Earth, what was thought to be empty space between the stars was found instead to harbor tenuous but vast clouds of gas and dust, and the starlight that we see was revealed to be the result of thermonuclear reactions that release energy in the process of converting nuclei of lighter elements into those of heavier ones. Paradoxically, during the same period the planetary sciences stagnated. Although our understanding of the universe grew rapidly during the first half of the century, knowledge of our own Solar System grew relatively slowly and, in some respects, became distorted during the first half of the century.

The most spectacular case of such warping of knowledge involved Mars and Percival Lowell's claims around the (last) turn of the century of an advanced intelligent civilization on the Red Planet. The press seized upon his statements and, along with Lowell's own efforts, largely convinced the general public that his claims were true. On the other hand, professional astronomers scoffed at such assertions, as they could find no supporting evidence. The press and radio became preoccupied with science fiction characters such as Buck Rogers, Flash Gordon, and their kind, and Orson

Welles's vivid radio invasion of Earth by death-dealing Martians scared the daylights out of millions of radio listeners. Such sensationalism exasperated most astronomers, and along with the fact that advances in our knowledge of the Solar System had become few and far between, led them to consider that field as essentially at a dead end and, indeed, faintly ludicrous. Planetary astronomy ceased to be an area of serious scientific study, but the public retained an intense but myopic and romanticized interest. Solar System study and research were ignored to the extent that, by the 1950s, there was only one professional planetary astrophysicist in the entire world—Gerard P. Kuiper of the University of Chicago's Yerkes Observatory.[15]

Before the space age, knowledge of the Solar System was for the most part obtained from direct visual observations, enhanced since Galileo's time with telescopes that, even by the time of Percival Lowell, remained essentially similar instruments. The introduction of photography and spectroscopy provided significant advances in astronomical research, but observations effectively remained confined to wavelengths of light visible to the naked eye.

But just what was known about the Solar System, and more to the point, its outer reaches, before the space age in general and the Voyagers in particular? Clearly, without such a milepost we can not hope to appreciate the advances in Solar System astronomy occasioned by the Voyager program. Elementary textbooks (there were no advanced ones on the subject), popular books, professional reviews of specialized areas of research, and the like, offer the best insight into what was known.

By far the most widely used elementary astronomy textbook for university students in the decades before the Voyager launch was Robert H. Baker's *Astronomy,* which went through many editions, with the sixth published in 1955, a few years before the creation of NASA. In it are tables listing the properties of the planets and their satellites as then known. As for the big outer planets (Jupiter, Saturn, Uranus, and Neptune), their orbits around the Sun were well known and, because all have satellites, so were their masses. In contrast, their diameters were less well known, the uncertainties growing as we progressed outward from Jupiter to Neptune, and the average densities had similar ambiguities. Those densities, however, were significantly less than those of any of the terrestrial planets. On none of the gas giants did we view a definite solid or liquid surface, only the top of an enveloping cloud cover. Tilts of the rotation axes and atmospheric rotation periods (how fast the clouds move) were known fairly well for Jupiter and Saturn, but were only approximate for Uranus and Neptune, which had never shown any definite features through the telescope.

Prominent cloud belts and spots of varied colors on Jupiter revealed numerous atmospheric currents (winds) and disturbances (storms), and rarer and more subdued markings on Saturn indicated a similarly active atmosphere. On both planets the atmospheres did not rotate as a whole but generally moved faster near the equator than elsewhere.

Uranus and Neptune had higher densities than either Jupiter or Saturn, but all four had sufficiently lower densities compared to the terrestrial planets that the two groups of bodies must differ greatly in chemical composition and internal structure. Although methane had been observed on all the giants, and ammonia on Jupiter at least, the chief constituent of all the giants was, presumably, hydrogen. A popular model of the interiors of these bodies featured a relatively small rocky core, a middle layer of compressed water ice, and an outer shell composed mostly of gaseous hydrogen.

Among the few specific features known, the Great Red Spot on Jupiter had been observed off and on for three centuries, but its nature was unknown, as was the reason for its often vivid color—or the varied colors of other Jovian cloud features, for that matter. Saturn, for its part, had a conspicuous ring system with at least four concentric ringlets that were all extremely thin and composed of myriad small solid particles; gaps between some of the rings were due to the gravitational influences of the planet's satellites.

About Uranus and Neptune little was known beyond descriptions of their orbits. The former appeared greenish to the eye, the latter a bluish gray, but no definite features had ever been observed on their dim disks. However, these bodies rotated in somewhat less than an Earth day, and Uranus's axis of rotation lay in the plane of its orbit around the Sun; it "lies on its side," so to speak. Both were apparently covered completely by clouds and had methane in their atmospheres; beyond that, nothing was known about them.

Many satellites orbited the giant planets, and some at least always kept the same face toward their primary, but for none were diameters, masses, and densities better than rough approximations. Similarly, no companion had revealed any definite surface features, and there were only hints as to their chemical compositions and internal structures.

Jupiter had twelve satellites, four of which were comparable to our Moon or Mercury in size and showed tiny disks in telescopes. Saturn had nine companions of which the largest, Titan, was orange in color and had an atmosphere containing methane. Another, Iapetus, had one side that is six times brighter than the other. The low densities of the ringed planet's inner moons suggested that they were composed mostly of different ices. Uranus had five satellites, none of them large, which moved in orbits that, like the planet itself, lay on their sides. Neptune had two attendants, of which one, Triton, was almost the size of Mercury.

That was it! Our knowledge of the Solar System could be neatly summed up in a mere seventeen pages of Baker's book.[16] Of course, more details could be found in technical publications, but they did not alter this basic and sparse picture much at all. One can see how glacially slow was the progress in our knowledge of the outer Solar System during these days by glancing at the next edition of Baker's book, which appeared in 1959. Among the advances mentioned were theoretical studies indicating that the giant planets were composed mainly of hydrogen, which changed from a gaseous to a liquid and then to a solid state with increasing depth. Another advance came from the way in which the star Sigma Arietis faded as Jupiter passed in front of (occulted) it, suggesting that the planet's atmosphere above the cloud tops was composed mainly of hydrogen, with most of the rest being helium, with only a whiff of other gases. In a similar vein, molecular hydrogen had been directly observed on Uranus. Finally, Jupiter's satellites Europa and Ganymede appeared to be covered with snow made of water ice. Again, that was it![17]

A more advanced textbook was *Astronomy: A Revision of Young's Manual of Astronomy* by Henry Norris Russell and coauthors, a two-volume set, the first volume of which was devoted to the Solar System. A revised edition of volume 1 came out in 1945, and, as Russell was then the undisputed dean of American astrophysicists, it is considered a benchmark work. However, despite forty-three pages devoted to the giant planets, there is little more definite information about the physical nature of the giant planets and their satellites than appears in Baker's works. So slow was the progress of planetary studies in midcentury that many researchers as late as the 1960s used Russell's twenty-year-old work as a useful reference.[18]

The few technical reviews of studies on planetary astronomy further underscored it as a "backwater" field. Perhaps the most illustrative example is a 1944 article by comet expert Nicholas T. Bobrovnikoff that appeared in the prestigious journal *Reviews of Modern Physics*. In it, he wrote, "The identification of methane and ammonia in the atmosphere [sic] of the major planets and of carbon dioxide in the atmosphere of Venus appears to have solved the planetary problem in its entirety."[19]

This smug assertion illustrates not only how meager was our store of knowledge of the planets at the time but also the widespread belief among scientists that our knowledge had progressed about as far as it could go in the field, and that, indeed, scientists already knew all that we needed or wanted to know about the Solar System. Decades later, after the initiation of NASA space programs, similar attitudes among a wide variety of "distinguished senior scientists" discouraged planetary research in general, and deterred investment in expensive space probes to the outer planets.

Fortunately, shortly after the end of World War II, Kuiper realized that the era of direct space exploration by means of rockets was fast approaching. That being the case, he believed that a comprehensive review of our "current" knowledge of the Solar System would provide an invaluable benchmark against which we could measure future advances in the field made by means of both spacecraft and ground-based observations.

Significantly, Kuiper secured major funding for this substantial effort from the U.S. Air Force (which was developing an interest in rocketry and space flight) rather than from a "scholarly" organization such as the National Science Foundation; NASA had not yet been invented, and its NACA predecessor was still largely concerned with the design of aircraft wings.

The first volume of Kuiper's series on planetary astronomy was devoted to our star, *The Sun,* and appeared in 1953. A second, concerning *The Earth as a Planet,* came out in 1954; a third, covering *Planets and Satellites* was published in 1961, and a fourth on the *Moon, Meteorites, and Comets* appeared in 1961. Unfortunately, a projected fifth volume on planets and the interplanetary medium never appeared, so the series lacks anything on important topics such as planetary spectra. Nevertheless, these books served their intended purpose well; just leafing through them provides a vivid reminder of how much we have learned since they were published. Typically, the studies contained tables of measurements of some planet's (or satellite's) brightnesses in different colors and at different positions in its orbit, and perhaps a mathematical treatment of what those measures might reveal, but little about the body's composition or physical structure. The meager content of hard facts about the outer Solar System emphasizes just how little we knew then and just how much we have learned since they appeared, in large measure due to the Voyager program.[20]

Of course we learned much about the outer Solar System between the mid-1960s and the Voyager launches, and most of that new knowledge came from "ground-based" (telescope, aircraft, and balloon) observations, many of which were funded by NASA. Indeed, that agency largely was responsible for a rebirth of planetary astronomy as a progressive scientific and scholarly field. Not only did NASA launch space probes but also it funded a wide variety of observational and theoretical research on Earth in an effort to ensure that its space missions would carry the most appropriate instruments possible and make the most useful and significant observations. Perhaps NASA's greatest effort was to do everything possible by means of relatively low cost ground-based observations, reserving for the much more expensive space missions only those tasks that could be performed in no other way. In large part it did that job very well indeed.

This is not the place to record all the discoveries made about the outer Solar System in the years just before the Voyager launches—some presenting new opportunities, some new problems—for that has been done in many places. However, we can mention a few results that directly affected those missions. Perhaps the most startling finding was that there is intense radio emission coming from the vicinity of Jupiter. There are spectacular noise bursts, so strong that they can be detected by amateur radio receivers and, surprisingly, occur only when the satellite Io is in a particular position in its orbit. Then, too, there is a steady radio component produced by the Jovian equivalent of Earth's Van Allen radiation belts. This latter finding revealed the peril of passing through these regions, a problem that, as we have seen, was accentuated by the results from Pioneer 10 and 11 missions. Io itself was shown to be surrounded by a softly glowing cloud of sodium and other elements, though where this came from nobody knew for sure at the time.

As for Saturn, terrestrial observers published disputed findings of additional satellites close to the planet and additional rings outside of the well-known examples. If real, the former represented new targets of opportunity and the latter increased the probability that a flyby probe would be demolished by collision with some relatively tiny orbiting particles.

Many of these discoveries meant additional real or potential work for everyone concerned in planning Voyager mission strategy, from both the engineering and scientific aspects. Of course, discoveries did not stop after the spacecraft hardware had been designed, built, and integrated to the point of no return. In fact, one of the most important lessons that can be drawn from the Voyager project is the way in which, after launch, those who conducted it were able to take advantage of new situations while avoiding previously unknown pitfalls. Indeed, the flexibility of these vastly complicated machines, and of those who controlled them, all the result of inspired and diligent effort, was a fundamental reason why the missions not only did not fail but succeeded far beyond reasonable expectations.[21]

There was, however, one intriguing possibility that posed a unique and very real problem to mission planners—would the Voyagers cause a catastrophe that might be called a "reverse Andromeda Strain"? That is, might a few stray bacteria, viruses, or even complex organic molecules from Earth (such as those from a sneeze, a fingerprint, or even a breath of wind) wreak unintended havoc on some body in the outer Solar System?[22]

Early in the space age there was widespread concern that material brought from Earth by space vehicles might disrupt systems of life, or even of "prebiological" organic molecules, on other celestial bodies, thereby destroying the very evidence of life or the origins of life that the probes were sent to look for. Later, such worries seemed pointless when direct exploration of the Moon, Venus, Mars, and Mercury showed not only no signs of life at all but also no evidence of any complex organic molecules. However, it was a different story in the outer Solar System, for methane was known to exist on the four giant planets and on the moon Titan, and subsequently other organic molecules had been detected in Jupiter's atmosphere. Clearly it would be wise to "quarantine" the Voyagers, or the outer planets, depending on one's point of view.

Sterilizing a spacecraft to the point of eliminating not only all life but also all complex organic molecules, without ruining it, is essentially impossible. Any effective method, such as high temperatures, chemicals, or ionizing radiation, also tends to damage or destroy electronic or mechanical parts. In programs such as Apollo and Mars Viking, NASA has taken extreme measures to avoid contamination. However, there is no danger of contamination if no object from Earth impacts another celestial body. *Avoidance* was the solution adopted for Voyager. Careful selection of trajectories gave great confidence that neither the Voyagers nor their last-stage booster rockets would impact any planet or known satellite. In retrospect, that strategy worked.[23]

Lift-off at Kennedy Space Center in August and September 1977 was, in many respects, the most critical moment in the history of the Voyager's Grand Tour. Yet, in truth, it was only a beginning. In terms of space exploration and astronomical science, and in terms of the work load of the Deep Space Network, lift-off began what Douglas J. Mudgway referred to as the "Voyager Era," a time when "the two Voyager spacecraft repeatedly astonished the world with a flood of dazzling science data and images as they transmitted their data from Jupiter, Saturn, and Uranus."[24]

At the time of lift-off Voyager's mission was MJS '77, to explore Jupiter and Saturn. The old Grand Tour program had not been resurrected; rather, a new Grand Tour evolved as Voyager 1 and 2 passed new milestones in planetary exploration. The first of those milestones would be Jupiter and its satellites. Voyager 1's Jupiter encounter began in 1978, followed in 1979 by that of Voyager 2. As the Voyagers progressed, so too did humankind's knowledge and understanding of the Solar System.

FIRST ENCOUNTERS

STRANGE NEW WORLDS

By December 1978 the spatial resolution on Jupiter's disk by Voyager 1's cameras was better than that possible with any terrestrial telescopes and, a month before closest approach on March 5, 1979, it was beginning to exceed that of any images from Pioneers 10 or 11. As the spacecraft closed in on the planet, pictures were taken in several colors to provide a continuous record of changes in the Jovian upper cloud layers. Meanwhile, on February 28, the probe passed into Jupiter's magnetosphere, where the motions of charged particles are controlled by the planet's magnetic field. In fact, there were a number of crossings within a few days, due to variations in the strength of the solar wind and the rotation of the planet's asymmetric magnetic field.[1]

More and more detail was revealed among the tortured clouds of Jupiter's atmosphere as the spacecraft neared the planet, but the real revelations, those related to Jupiter's large Galilean satellites, came before closest approach. The most spectacular discoveries of Voyager 1's Jupiter encounter, and many would conclude of the entire Voyager program, was the bizarre nature of Io, the planet's closest large satellite.[2] Overall, this

Io, taken by Voyager 1 on March 3, 1979, against the
Jovian atmosphere. (NASA photo PIA-00378)

moon's true color is more nearly yellow than orange, with darker polar
regions and irregularly shaped whitish patches near the equator, and the
surface is strewn with a bewildering assortment of relatively small light
and dark features showing a wide variety of shapes and colors. At first,
scientists suspected that the "spots" might be impact craters, but later images made with greater resolution (Voyager 1 was roughly 20,000 kilometers from Io at closest approach) found not a single impact crater on
the entire satellite, even though pits as small as 1 kilometer in diameter
should have been visible. If the rate of crater formation at Io is similar to
that observed in the inner Solar System, some process must be obliterat-
ing them rapidly by geologic standards, completely resurfacing that moon
in one million years or less—and, perhaps much, much less.

High-resolution images indicated that the resurfacing agent is vulcanism, revealing over one hundred depressions similar to, but larger than,
calderas such as that of Kilauea on the island of Hawaii. Also present are
vast smooth plains and lava flows of various kinds, suggesting that the
erupted magmas varied from very fluid and free flowing to highly viscous
and stiff. However, in contrast to Earth and Mars, the calderas are not associated with much vertical relief; in other words, there are no tall volcanic mountains. Io also lacks mountain ranges such as the Andes or
Rockies, but it does have isolated, jagged peaks that tower as much as 10
kilometers above their surroundings and so must be composed of solid

rock with substantial strength. Their general appearance resembles that of broken, partially foundered fragments of an earlier, rocky crust. In addition, the satellite exhibits other indications of tectonic activity such as faults, scarps, and graben (regions of sunken crust bounded by fault lines).

Evidently Io is an active world geologically, but just how incredibly active it is was revealed by a photograph taken on March 8, soon after Voyager 1's closest Jupiter approach and at a range of 4.5 million miles. At such a relatively large distance not much surface detail was expected and instead the photograph was taken for navigational purposes. As part of this effort the original image was digitally processed to bring out two reference stars, but it also showed something else. Linda A. Morabito, a Navigation Team member, noticed a faint, thin crescent just above the satellite's eastern limb. The first impression was that of a previously unknown celestial body lying beyond Io and partially peeking out from behind the satellite. However, a such an object would have been discovered long ago by terrestrial telescopes, and so the crescent was more reasonably identified as a thin, umbrella-shaped cloud of gas and/or dust rising some 270 kilometers above Io's surface.

There was another odd feature on the same image: a bright spot just inside the night side of Io. This was another eruptive cloud, one extending high above the dark surface and into a region where the Sun was shining. The position of each plume was above a surface feature that previously had been identified as volcanic. Io had two volcanos in vigorous activity at the same time!

Once the members of the Imaging Team realized what was going on, they eagerly searched other Io images for evidence of eruptive plumes from additional active volcanos, and a total of eight, later amended to nine, were found (with more confirmed later by subsequent probes). Some plumes are dark when seen projected against the moon's surface, which means that the erupted material has a low reflectivity; some plumes are symmetric, some are not.

The density of Io's atmosphere is so low (it may be mostly sulfur dioxide, SO_2) that erupted material is essentially unaffected by air resistance and follows a ballistic path, like an artillery shell. Observed plume heights indicate ejection velocities of up to 1 kilometer per second, and the halos and rings seen around vents correspond to areas where the expelled material falls back to the surface.

As Io was probably formed some 4.5 billion years ago, what powers its vulcanism today? The satellite is so small that it should have lost most of any original internal heat unless it contains an improbably high proportion of long-lived radioactive elements. So what is the cause? By one of those coincidences that happens every so often in astronomy, the answer (which was actually a prediction of vulcanism on Io) was published just before Voyager 1's flyby. The energy comes from tides raised in the body of the Moon by Jupiter's gravity. Like the other Galilean satellites, Io always turns the same face toward Jupiter as it orbits the planet, whose gravitational tidal forces deform the moon, elongating it in the direction of its primary. If that were all, this "tidal bulge" would always point toward the center of Jupiter and that would be that. However, the satellites Europa and Ganymede continuously deform Io's orbit, and one result is that the tidal bulge does not always point exactly at the planet. In response, the satellite changes its shape so that the bulge does face Jupiter. This constant flexing heats the body of the satellite much as an automobile tire is heated by the flexing it undergoes during a long, high-speed drive.[3]

The energy to heat Io's interior is essentially unlimited, as it comes from the enormous reservoirs of Jupiter's rotational energy and the orbital energies of the satellites involved. But just what are the eruptive plumes and lava flows made of? Over the aeons since Jupiter's system formed, it seems inevitable that light molecules such as water and carbon dioxide, let alone those of substances such as hydrogen, helium, and nitrogen, have long ago escaped from the moon's feeble gravity. (Moreover, its average density is similar to that of common terrestrial rocks, leaving little room for very light constituents.) On the other hand, Voyager infrared measurements indicated surface temperatures, even in localized "hot spots," of only 290 degrees K, far too cold for lavas similar to those on Earth. So what was left?

Sulfur was and is the best candidate. It is heavy enough for some to have survived on Io, yet is relatively volatile, that is, it has low boiling and evaporation points. In addition, elemental sulfur has a bewildering variety of so-called allotropic forms, which exhibit all sorts of colors and physical properties. Additional players might well be sulfur compounds such as sulfur dioxide and hydrogen sulfide (H_2S), either in gaseous or solid form. The fact that there is emission from ionized sulfur atoms in the vicinity of Io's orbit strengthens the identification—the ions are just blown off the satellite.

Thus, a picture emerged of Io as a sort of "low-temperature Earth" (but not quite, as we will see below) with sulfur and its compounds taking the place of Earth's more familiar molten rock and water vapor (though sulfur and its compounds are common, if minor, constituents of terrestrial volcanic eruptions). However, scientists realized that silicates or silicates enriched in sulfur could also play a role. The satellite's episodic eruptions neatly explained the disparate temperatures obtained earlier by infrared telescopic observations from Earth. The Imaging Team responsible for monitoring Voyager's encounter with Io, Voyager engineers, and associated scientists indicated that they were more intrigued by the Io discoveries than by almost any other phase of the Voyager discoveries. That excitement and interest, they suggested, was shared by many in the general public.[4]

Io had another big surprise in store decades later. Though the main thrust of this work is to "tell things as they were," there has to be an exception here. On Thanksgiving Day 1999 NASA's Galileo probe made a close pass by the moon and revealed an amazing sight. Near the satellite's north pole was a huge "curtain of fire," similar to the linear chains of lava fountains that occur along "rift zones" on the island of Hawaii, but vaster by far. The Io display, dubbed Tvashtar after the thunderbolt-forging Indian god, was 25 kilometers long and more than a kilometer high; in fact, it was observed telescopically from Earth in the infrared! The most surprising property of this eruption was its high temperature, probably about 1,600 degrees K, which indicated silicate volcanism very much like that on Earth, and much more violent than anything directly observed by the Voyagers. It seems that Io still has surprises in store.[5]

Europa, the next satellite outward from Io, was known to have a rather low density, consistent with water—probably in the form of ice—being a significant part of its makeup. Voyager 1's closest approach was far more distant than for any of the other Galilean satellites, but as the smallest of the quartet, it seemed least likely to hold any big surprises. Initial images were unimpressive, showing a more or less whitish body with only subtle variations in surface color and reflectivity. Better photographs revealed

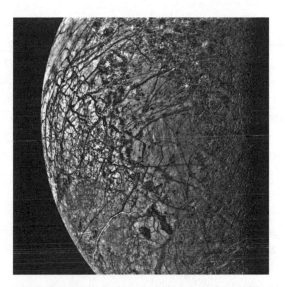

Image of Europa taken by Voyager 2 on July 10, 1979.
(NASA photo PIA-01503)

only a few features that resembled impact craters, and even these had low
vertical relief and lacked "rays" such as are seen on, for example, the
Moon; neither were there mountains or canyons. In fact, Europa appeared
to be the smoothest and "flattest" body in the entire Solar System.

However, the most startling aspect of Europa's surface proved to be a
complex pattern of arc-like linear features up to 200 kilometers wide and
several thousand kilometers long. On the whole, this moon looks like a
hard-boiled egg whose shell has been cracked time and time again, and it
bears an eerie resemblance to the canal-laced drawings of Mars made by
Percival Lowell almost a century earlier.

The tidal heating mechanism that explains Io's volcanic activity should
also operate to a lesser extent on more distant Europa. It could provide
enough energy to keep some of the satellite's abundant water in molten
form and provide a "moon-wide" liquid ocean that would be covered by
a thin shell of solid water (that is, ice) frozen by its exposure to the cold
of space so far from the Sun. The "cracks" would be caused by currents
in the underlying sea that moved the overlying ice floes around, fracturing them in the process. Liquid water would be exposed in the cracks for

a time but would soon freeze, though perhaps not without leaving some evidence of its former presence. Probably there would be some "geysering" during the cracks' open stages, and the ejected material would quickly freeze and fall back to the surface, covering whatever was there with a blanket of snow, perhaps adding some slight coloration. In addition, the relatively thin ice crust would "cold flow" like a glacier and obliterate any trace of an impact crater in a short time by geologic standards. Thus the crust of Europa that we see today probably is relatively young, though older than that of Io. There was considerable anticipation of what Voyager 2 photographs would show from closer up, but in any case Voyager 1 had revealed another unique world, one unlike any ever imagined.[6]

The first, most distant images of Ganymede showed a surface that resembled the Moon with fairly well defined darker and lighter areas, as well as bright spots that resembled rayed impact craters. Higher-resolution pictures, however, showed a body that was very different from our own satellite. The dark areas have sharply defined edges, are shaped like various polygons (triangles, rectangles, diamonds, pentagons, hexagons, and the like), and are almost saturated with impact craters; that is, the formation of any new craters would wipe out enough old ones so that the observed number would not change substantially. Because of the large crater density, these regions must be relatively old, probably similar in age to our Moon's (bright!) highlands, say, four billion years or so; presumably they are remnants of an ancient crust that once covered the entire body. An old age for the dark areas is supported by the fact that most of the craters in it are relatively shallow. Presumably this is because the surface material contains a large proportion of water ice that, in spite of the low temperatures, slowly flows and levels out elevation differences, much as glaciers flow downhill on Earth. The bright rays are probably lighter-colored material that was blasted out from beneath the surface.

Separating the darker areas are more enigmatic features: bright stripes that crisscross the surface in a wide variety of patterns ranging up to 100 kilometers in width and 1,000 in length. These stripes are made up of sets of parallel grooves, each no more than a few hundred meters deep and resembling parts of huge "contour-plowed" fields. The sets never overlap, for one group ends where it meets another, but some are bent or otherwise contorted, and some have parts that are displaced along what appear to be fault lines. The surface density of impact craters on the grooved terrain is much less than that on the darker areas on Ganymede but still about the same as that of lunar maria and the oldest Martian plains. This suggests that whatever process formed the grooved terrain stopped operating some three billion years ago.

The most probable interpretation of the Voyager 1 images is that early tectonic activity sundered Ganymede's original crust (the dark polygons), some of which then sunk. There were repeated upwellings of material (water and water ice were probably major constituents) between the remaining chunks of old surface, and the entire assemblage repeatedly was pushed around in various directions by subsurface currents. However, because the tidal heating mechanism generates relatively little heat, this small world froze solid a few billion years after its formation, and little has happened to it since except for occasional impacts. Supporting the view that there is little new on Ganymede is its surface's overall low relief; there are no deep impact basins or high mountains of any kind. Evidently, over billions of years the slow creep of the ice-rich crust has smoothed out any original height differences.

Callisto is the darkest of the Galilean satellites and has the lowest average density, implying a large proportion of water ice in its makeup. Voyager 1 images showed that most of the surface is almost saturated with impact craters, and thus may date back four billion years or more. Some of the craters are surrounded by brighter material evidently excavated from below the surface by the impacts, but nowhere on this moon are there any large height differences. However, the largest and most spectacular features are two huge multiring structures, each covering a significant part of the moon's surface. These "bull's eyes" resemble the surface of a tranquil pond just after a stone has been thrown in, only in Callisto's case the expanding "waves" have been frozen in place. There are similar structures on our Moon and the planet Mercury, but on Jupiter's satellite there are no large depressed basins at the centers, but only brighter, more lightly cratered splotches. Evidently the large impacts that formed these structures occurred after most of the present crust was formed, but still early enough in the history of the Solar System for the actual craters be eliminated by cold flow of the icy crust, leaving only "palimpsests" behind. Callisto is, apparently, a frozen world on which

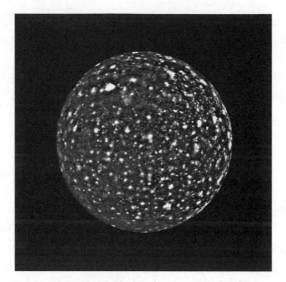

Callisto, as viewed by Voyager 1 on March 6, 1979.
(NASA photo PIA-02253)

little, or nothing, has happened since the early days of its existence except for a gradually decreasing barrage of impacts.

Of Jupiter's many small satellites, only Amalthea, the closest of them all, passed within effective range of Voyager 1's cameras. This moon is only a point of light as seen from Earth, though astronomers had determined that it is dark, reflecting only about 5 percent of the sunlight that falls on it, and reddish in color. There was considerable interest in what spacecraft images would show, as this satellite's size lies between those of Mars' tiny attendants Deimos and Phobos on the one hand, and the Galilean moons on the other. The best pictures were taken shortly after closest approach to Jupiter, and revealed an elongated, potato-shaped body roughly 165 by 140 kilometers, with the longest axis pointed toward Jupiter. The majority of its surface is indeed dark and reddish, but there are several brighter, whitish spots of unknown nature. Either Amalthea was formed as an oddly shaped object or, more probably, was deformed by past impacts that blasted portions of it away. The satellite's weak gravity has been unable to shape it into a spherical form, which shows that the moon's body has some structural strength and is not, for example, a loose agglomeration of gravel. In the light of Io's violent volcanic activity, it is possible that Amalthea's red color is due to a covering of material erupted from the former satellite.

Another of Voyager 1's exciting revelations had to do neither with the planet itself nor with its known or suspected satellites. A long-exposure (a bit over eleven minutes) image taken shortly before closest approach to Jupiter, as the spacecraft was crossing the planet's equatorial plane, showed it to be surrounded by a ring. This feature shows no obvious sub-rings or brightness gradient toward Jupiter and was less than 30 kilometers thick (and probably much less than that). The ring's surface brightness is very low, and would be invisible to the naked eye if viewed face on; the Voyager detection was aided by an almost edge-on presentation.

The ring's outer edge was some about 128,000 kilometers from Jupiter's center, 1.8 times the planet's radius and about halfway out from the upper cloud surface toward Amalthea. Interestingly, data on charged particles observed by Pioneer 11 (namely, a lack of them) had indicated much earlier that "something," either an undiscovered satellite or ring, was lurking at just that distance from Jupiter.[7] Moreover, though the ring had not been detected by astronomers on Earth, its shadow possibly was. Sometimes a thin, faint, darkish line, known as the Equatorial Band, can be seen running through the center of Jupiter's bright equatorial zone, never appearing more than a few degrees or so from the planet's equator; this might have been evidence of a ring oriented almost edge-on to the Sun.[8]

As Voyager 1 neared Jupiter, scientists looked forward to a flood of information about the planet's colorful and ever-changing cloud patterns, not to mention myriad spectacular images. Telescopic studies over the years had shown that the giant planet's upper atmosphere contains strong and variable winds that depend to a large extent on latitude, atmospheric disturbances—both light and dark in color—that come and go on various time scales, and at least one feature, the Great Red Spot, that has endured for centuries.

Jupiter's weather was known to differ from our own. For one thing, the intensity of Jovian sunlight is only 4 percent of that at Earth, and the flow of internal heat should be much greater than on our planet and is the primary driver of atmospheric motions. This giant planet has no solid or liquid surface, in contrast to Earth and the other terrestrial planets; things just become denser and denser without any noticeable "change of phase." As a result, there are no complications due to the presence of distinct oceans and continents and, by the same token, no mountains to complicate things; neither are there any appreciable seasonal effects, for the planet's axis of rotation is nearly perpendicular to its orbital plane. Various weather belts depending on latitude are more pronounced than on Earth (though we do have them: the doldrums, the trade winds, and the roaring forties are examples), and the planet's very rapid rotation produces strong Coriolis forces, like those that cause the rotation of our hurricanes and high- and low-pressure areas. Then too, although terrestrial clouds are limited to liquid and solid particles of H_2O (and, in some cases, dust), there was strong evidence that, in the case of this giant planet, condensates of several different compounds formed different cloud layers at different levels in the atmosphere. One of the most puzzling features of the planet's upper layers is the fact that its Southern Hemisphere exhibits a much greater variety of phenomena than the Northern, the Great Red Spot and the South Tropical Disturbance being two of the most prominent examples. Why there is this difference between the two halves of Jupiter was, and is, a mystery. In sum, Jovian weather is, in various aspects, both simpler and more complicated than Earth's, but it definitely is different.[9]

One big surprise concerning Jupiter's cloud structure came from Voyager 1's infrared observations. Previous terrestrial research had shown that the planet's "white" zones are colder than the darker belts, presumably because the former areas are at higher altitudes than the latter (the Great red Spot is an exception, as its top is relatively cold). Further, certain relatively small bluish regions near the planet's equator show the highest temperatures of all. These areas are clear of upper-level clouds and owe their tint to the same mechanism that produces the blue color of Earth's cloudless daylight sky, namely, scattering of sunlight by molecules of gas. Through these clear patches we can see deep into Jupiter's atmosphere, where the temperature is high because of the enormous amount of internal heat still left over from the planet's formation.

Every time terrestrial observers improved their instruments so that they could observe smaller regions on Jupiter, they found higher temperatures, up to a maximum of 260 degrees K. On the basis of this work, Voyager scientists expected to find regions, too small to be investigated from Earth, where the temperatures were even higher—in other words, holes that were clear down to very deep levels. However, no temperatures above 260 degrees K were observed. Evidently there is, or was at the time, some sort of opaque "floor" beneath which we cannot see, even in the infrared. This layer may be composed of clouds, or it may be some sort of molecule that absorbs light over a wide range of wavelengths.

Infrared observations also provided an estimate of the amount of helium in Jupiter's atmosphere, a quantity that has far-reaching implications. For example, in current astrophysical thinking the newly formed Sun was composed mostly of hydrogen, with most of the rest being helium, and only a whiff of all the other elements. Just how much helium was there at the start determines a number of things such as how much brighter our star has gotten since then and how long it will continue to shine.

Unfortunately, it is difficult to estimate that initial abundance. The Sun itself is of little help because it shines by fusing nuclei of hydrogen into those of helium, and some of that "manufactured" helium may be contaminating our star's outermost layers, which are the only ones we can observe. In any case, the amount of "new" helium produced depends on what was there in the first place. Moreover, there are problems with current methods of determining accurately the current abundance of helium in the Sun's outer layers. Earth is not of any help whatever, as much of whatever helium it once had has escaped into space because our planet's relatively feeble gravity cannot restrain the gas's low-mass molecules.

Jupiter, however, has no nuclear reactions in its interior, and its strong gravity means that it still retains pretty much all the material that it started with—even hydrogen, the lightest element of all. Thus, assuming that the planet's helium is evenly mixed throughout, its abundance in the upper atmosphere should be that of the primordial Sun. Even Voyager data required a rather sophisticated analysis to arrive at the desired value, but the result is that about 11 percent of the volume of Jupiter's atmosphere is helium.

Other infrared observations revealed that the presumably higher clouds in the bright zones have only a small effect on the variation of temperature with depth—a mild surprise. There was also confirmation of some curious asymmetries first suggested by observations from Earth. Jupiter's tropopause, or transition zone in the atmosphere above which lies the stratosphere, is lowest and warmest near the equator, becoming higher and cooler toward the poles; however, the warmest area is not exactly on the equator but some fifteen degrees in latitude south of it. In contrast, the upper stratosphere near the poles is colder in the Southern than in the Northern Hemisphere. The cause of these differences, like the north-south asymmetry of cloud features mentioned earlier, is a puzzle.

Infrared observations of the Great Red Spot, though made with spatial resolution far better than any attained before, nevertheless were blocked by clouds at a relatively cold, and thus presumably high, altitude, as were earlier terrestrial ones. The spot had long been known to rotate as an anticyclone (high-pressure area), that is, counterclockwise when seen from above, just like "highs" in Earth's Southern Hemisphere, making one "revolution" in about six days. This behavior requires a warm "core" of which no trace was detected; any such region must lie below the level accessible to Voyager 1's instruments. Yet rising currents there have to be, for high-resolution images show the interior of the Great Red Spot to be filled with the puffy cloud tops of ascending columns of air—gigantic cumulus clouds.

Direct closeup images of Jupiter provided some additional surprises. For one, the pattern of alternating zones and belts with different east-west winds, long known at lower latitudes, was found to extend into the polar regions. This contradicted the previously held notion that atmospheric flows at high latitudes were dominated by up and down motions, much like a field of cumulus clouds on Earth.

Another important result, but one that did not receive much publicity in the mass media, was that features in the same region but of greatly different sizes have similar velocities. This property indicates that real motions of air masses are taking place, and not just wave motions (such as when a laid-to ship experiences waves passing beneath it, but the water itself does not move much with respect to the vessel).

On another topic, astronomers had long observed bright, white spots that often appear suddenly in Jupiter's equatorial regions, become drawn out into long wisps, and then fade away. Voyager images proved that these features are indeed just what they look like, namely, ascending columns of air whose cloudy tops are caught by fast upper-level currents and elongated downwind. On Earth, energetic cumulonimbus clouds that grow high enough have their tops sheared off by stratospheric winds and exhibit "anvil heads" that may stretch for hundreds of kilometers, and evidently the same thing happens on Jupiter. However, the individual Jovian "cumulus clouds" are hundreds, instead of a few kilometers in diameter, and the associated anvil heads may extend thousands of kilometers from their places of origin. As expected, the centers of these spots show lumpy shapes resembling cauliflower heads, just like towering, anvil-forming cumulus clouds seen in images taken from Earth satellites.[10]

One aspect of Jupiter's appearance that often causes puzzlement is that the cloud features do not look anything like terrestrial ones. Convoluted complexes may yet have sharply defined boundaries that look positively unreal, an appearance that resembles the initial stages in the mixing of two different colors of oil-based paint more than anything that possibly could happen in an atmosphere. Part of the explanation lies in the immense size of the planet compared to Earth, for boundaries that appear perfectly sharp in many Voyager images may actually be "fuzzy" on scales of tens or even hundreds of kilometers. Similarly, those who have spent time in mountains will have noticed that what appears to be a sharp-edged cloud when seen from far away actually appears as a sort of indistinctly bounded fog when seen from close up. Although there are complex Jovian atmospheric features with very sharp boundaries, the highest-resolution Voyager images show that many of the planet's clouds actually do look like "real" clouds when seen with sufficient resolution.

Seen in detail for the first time were interactions of smaller features with larger ones such as the Great Red Spot and the long-lived "white ovals," as well as between the smaller but more common ones themselves. The larger features possess enough integrity to survive encounters essentially intact, but all sorts of things can happen to smaller ones: they may merge, distort, or apparently remain unchanged.

As Jupiter, like the other giant planets, has no definite solid (or liquid) surface, it is not obvious what "wind speed" means on such a body. In other words, what is the reference that atmospheric velocity is measured against? Historically, telescopic observers have compared speeds of different features with respect to each other, deriving relative speeds. Starting in the 1950s, radio astronomers detected emissions from Jupiter that varied periodically. Eventually it became clear that this repetition rate was the rotational period of the of the planet's massive interior, and thus it provided a reference for determining objective atmospheric velocities.

In general, wind speeds on Jupiter depend on latitude, and are arranged in planet-girdling belts parallel to the equator, some of them relatively narrow. The major currents appear to have persisted for at least a century, despite the changing cloud patterns. In many cases the transition from one wind regime to another is rather sharp, and Voyager 1 images provided the first closeup views of the resulting complex cloud patterns at such boundaries.

This is but a small sampling of the meteorological results from Voyager 1's flyby of Jupiter. Not only was the best information far more detailed than anything attainable from Earth, but there were several months of continuous observation when that condition held. All in all, the atmospheric scientists on the Voyager teams were the lucky recipients of an enormous amount of new information, the likes of which had not been seen since the 1960s, when meteorologists first began to receive data about Earth's weather from orbiting satellites. They were kids let loose in a candy store with no one watching them.

As there are tremendous cumulus clouds on Jupiter, should there be lightning? As fights still rage as to the details of how lightning is produced on Earth, there were differing opinions on whether the phenomenon occurs on the giant planet and, if so, whether it could be observed by the Voyagers. Images of Jupiter's night side settled that issue, for they clearly showed energetic lightning bolts playing among the planet's cloud tops. The flashes bunch in clusters and occur more or less uniformly over the planet. Moreover, radio receivers aboard the spacecraft detected "whistlers," very-low-frequency radio waves generated by lightning in Jupiter's polar regions. Appropriately, typical Jovian thunderbolts are titanic by terrestrial standards, comparable in energy to rare "superbolts" on Earth, and provide another interesting case of mythology "predicting" reality.[11]

Images of Jupiter's night side clearly revealed the presence of polar aurora in three or more layers above the cloud tops. Like their terrestrial counterparts, these Jovian glows can vary in minutes or even seconds, and they are bright—one could drive an automobile without headlights (but carefully!) using just their illumination.

Voyager 1's measurements of the strengths and directions of magnetic fields, along with the energies and abundances of various charged particles (electrons, sulfur ions, oxygen ions, and so forth) added an enormous amount of detail about the incredibly complex and variable phenomena that occur in Jupiter's magnetosphere. Along with the planet's strong, "off axis" (with respect to the axis of rotation) magnetic field and associated intense radiation belts, material ejected from Io by one means or another at more or less random times complicates the situation even more.

As Voyager 1 receded from Jupiter and headed for Saturn, it was in remarkably good condition. In fact, it had only three problems, and one of them self-corrected. First, the photopolarimeter's "analyzer wheel," which held a set of filters to isolate different spectral regions, failed to work correctly, precluding better estimates of cloud-top heights and the sizes and shapes of cloud particles. Second, during passage through the most intense region of the radiation belts, the flight data system's time reference shifted some eight seconds, causing smearing of some of the Io and Ganymede images. Finally, the ultraviolet radiation instrument became inoperative during the Jupiter encounter but annealed after leaving Jupiter's magnetic field. All in all, the probe was in fine shape.

Now attention shifted to Voyager 2 and what it might reveal. The probe's closest approach was scheduled for July 9, 1979, and observations from Earth and both Voyagers indicated that the Jovian planet's face would not appear drastically different—the Great Red Spot would still be there, for example. However, there was a great deal of speculation as to what Io would be doing, as well as anticipation of what images of Europa with much better spatial resolution than any taken by Voyager 1 would reveal, along with the chance to photograph regions of Ganymede and Callisto that the previous probe did not view.[12]

Six of the eight active volcanic regions observed by Voyager 1 were still going strong, though the largest of all had apparently subsided, and no new eruption plumes were seen.[13] There were several large-scale changes in the appearance of the satellite's surface, suggesting to astronomers that resurfacing, at least in the plume regions, was fairly rapid, perhaps depositing a centimeter of material every century—or even much more.

Europa revealed two types of terrain; darker, mottled areas and bright regions crossed by numerous dark lines and bright, narrow ridges. None of these features have much vertical extent, and the new photos strengthened the impression that we are seeing a relatively thin solid crust of ice that floats on, and has been ruptured by, an underlying ocean of water that completely covers the rocky silicate body of the moon. The better pictures also revealed three impact craters—but only three—indicating a surface much older than Io's, but still perhaps only a hundred million years old, more or less.

On Ganymede, the bright grooved terrain showed itself to be even more complex than Voyager 1 indicated. The largest area of dark, ancient surface shows a pattern of thin, concentric bright arcs, closely resembling part of one of the giant bull's eye target features on Callisto. On the old crust are a number of palimpsests or circular bright spots with little or no vertical relief. Presumably these are the only remaining traces of ancient impact craters that otherwise have been erased by cold flow of the icy crust over billions of years. Voyager also found smooth younger areas where the grooves have probably been covered by water that then froze, as well as weird-looking regions where short sections of ridges and grooves tend to butt up against each other at right angles. Judging by the density of impact craters found in different regions, it appears that the dark areas on Ganymede date back to the moon's earliest days. On the other hand, some of the bright areas were formed over billions of years, although none appears to be younger than several billion years. Apparently the gradual thickening of the satellite's rigid crust put a stop to extensive tectonic activity some time ago.

The previously unseen side of Callisto turns out to be even more heavily cratered than the one imaged by Voyager 1, and clearly almost the entire surface is very old. Moreover, there were no new features like the two bull's eyes found by that preceding probe. In short, Callisto has been a geologically dead world for a long time, with only impacts causing any changes.

Jupiter's ring, which appeared essentially featureless to Voyager 1, turned out to have some structure after all. There is a relatively bright segment, or annulus, where the particle density is highest, that is only about 800 kilometers in radial extent. Both inside and outside of this ringlet is fainter-appearing material some 5,200 kilometers in width. Interestingly, there are traces of faint ring material that appears to extend right down

Table 9.1

Voyager Closest Approaches (Kilometers from Center of Mass)

Body	Voyager 1	Voyager 2
Jupiter	348,890	721,670
Amalthea	420,200	558,370
Io	20,570	1,129,900
Europa	733,760	205,720
Ganymede	114,710	62,130
Callisto	126,400	214,930

Source: Data from E. C. Stone and A. L. Lane, "Voyager 2 at Jupiter," *Science,* November 23, 1979, 926.

to the top of Jupiter's atmosphere. If so, this may be an indication that ring membership is not permanent, with ring material lost to the planet being replaced from outside. In other words, the Jovian ring is not "original equipment." Candidates for supplying this ring material are the inner satellites, particularly Io, for obvious reasons. There were also three new candidates; all are tiny moons (no more than 50 kilometers or so in size) and orbit inside Io.[14] Metis and Adrastea are closer in than Amalthea and near the ring's outer edge, and Thebe lies between Amalthea and Io.

We should point out here that flyby probes such as the Voyagers are of little or no help in finding new *distant* satellites. Such spacecraft are unlikely to pass anywhere near these bodies, which must be small (to remain undetected so far), and there is no way of knowing what direction to look for them. Given this circumstance, the satellite discoveries achieved by virtue of the Voyager flights were remarkable. (Steve Synnott, one of the Navigation Team members, is credited with detecting more of the Voyager satellite discoveries than any other person.)[15] As regards Jupiter itself, Voyager 2 found that the major atmospheric currents mostly remained unchanged from the earlier flyby, even though many of the cloud features were different, confirming that these currents are long lived phenomena.[16]

Voyager 2 completed the principal tasks of its Jupiter encounter on August 5, 1979. Afterward it was "powered down" as it steadily followed its twin outward toward the next major targets: Saturn, its rings, and its moons. The stage was set for the next act in this unprecedented interplanetary drama.[17]

Voyager's encounter with Jupiter and the Galilean satellites marked the beginning of what would become a Grand Tour of the outer planets. It marked, too, the beginning of a new era of planetary discoveries and of a growing public awareness of Earth's position within the Solar System and its relationship to the planets and other celestial bodies. For the better part of the next three decades the Voyagers sent back to Earth new revelations and new discoveries, and it evoked new questions about our Solar System. The Voyager explorations became a unique and extended learning experience.

"Over the years," commented Deputy Project Scientist Ellis Miner, "the press, particularly those who were covering Voyager, became far more adept at accurately reporting the results that were obtained. I can remember when we had some press conferences prior to Voyager when there were often mistakes [caused by] an attempt to make the results seem even more spectacular. But Voyager results were so spectacular in their own right that they did not need to be embellished by the press—and seldom were."[18]

10

SATURN'S SYSTEM

ICE WORLDS AND RINGS

As Voyagers 1 and 2 left Jupiter and headed farther out into the Solar System, the gravitational assist provided by that planet meant that the "coast time" to Saturn would be only seventeen to twenty-five months for the two spacecraft. But the environment of space began to change markedly, for the intensity of sunlight at the ringed planet is four times weaker than at Jupiter and one hundred times weaker than at Earth, and spacecraft temperatures declined steadily creating greater possibilities for equipment failures. In theory, the radioisotope thermoelectric generators carried by the probes ensured that the spacecraft would have enough power at these greater Sun distances, but the draw on those power resources grew greater as heater circuits were used more and more heavily. Moreover, images required much longer exposures, making it more difficult to compensate for "smearing" due to the probe's motion with respect to a planet or its satellites.

Spectral absorption bands of methane are stronger on Saturn than on Jupiter, indicating that one looks deeper into its atmosphere before encountering a deck or decks of clouds that constitute the visible "surface." The increased scattering of light by molecules in a deeper atmosphere tends to obscure lower lying features, just as very distant mountains on Earth look indistinct even on the clearest days, so only the most towering cloud structures are visible.[1]

Voyager 1's encounter with Saturn began in November 1979, and Voyager 2's studies began in June 1981. The environment in the vicinity of the Saturnian system presented new technical challenges and problems, as well as new opportunities for discovery, for much less was known about Saturn and its satellites than about Jupiter and its moons. Although the prominent rings had been observed since Galileo's time, observable cloud features were rarely spotted, making it difficult to identify the various atmospheric currents that presumably exist. The planet itself has an average density less than that of water, indicating that its internal structure differs somewhat from that of Jupiter, although the chemical composition was expected to be mostly hydrogen, the rest being primarily helium with a trace of other elements.[2]

Perhaps one of the most outstanding things about Saturn was the lack of knowledge about that planet among the general public beyond the name and, among some, the perception of rings. Voyager changed that. Indicative of the change, the television and radio media, science writers, and scientists eagerly monitored Voyagers' Saturn encounters. Almost from the beginning, Ed Stone had devised an effective process that would help the media answer the questions they wanted answered. The Jet Propulsion Lab had learned to effectively structure the science activities during encounters so that "we could engage the public in the same thing we were sensing." "How do we tell the story to the Press so they can write about it?" If the story is too technical, it will be ignored. Thus, as we have seen, each press conference was preceded by what Stone described as a "daily activity" or conference wherein team leaders, public information officers, and managers discussed the most recent product from Voyager and devised a presentation that might best engage the attention and understanding of the media and the public.[3]

Concurrently, the media, and especially television, prepared to transmit Voyagers' discoveries to the world. Rex Ridenoure recalled, for example, that the media's "up-link" trucks had begun to assemble a week prior to the encounters, and that in time, the trucks and media were coming on site fully a month in advance. "It looked like the whole world was there."

Jancis Martin conducts a science briefing with the press. (Courtesy of Charles Kohlhase)

And, in a real sense, it was. There were media teams from New York City, Washington, D.C., Canada, South America, London, Tokyo, Paris, and elsewhere. Charles Kohlhase's Mission Planning Team was instrumental in both informing and educating the media about the missions' purposes and background, and the entire Voyager science community, working under the aura of Ed Stone, labored assiduously to communicate and explain the significance of incoming Voyager science data.[4]

Although serious scientific study of the Solar System was renewed in the 1960s and 1970s, observers on Earth found Saturn's system more difficult to study than Jupiter's. Because of the former's greater distance from the Sun, its surface brightness, as well as the brightness of its rings, was relatively low. At the same time, available detectors for all wavelengths of light were still relatively insensitive, so long exposures, low spectral resolution, poor spatial resolution, and "noisy" data were the norm. Big telescopes and large amounts of time on them helped, but both were hard to come by and expensive. As for Saturn's satellites, they were mere points of light. Hyron Spinrad, an observational astronomer at the University of California at Berkeley, characterized the situation aptly during one long, cold night at McDonald Observatory. "The outer Solar System," he sighed, "is really tough."[5]

In the two decades before the Voyager encounters, astronomers had added a substantial amount of additional information to what might be called our "classical" knowledge of the Saturnian system, but still the total was not great. The satellites provide a typical example. Estimates of their masses depended on the mutual gravitational effects on their orbits and were merely rough estimates. Their sizes were even more uncertain, as values depended on how efficiently their surfaces reflected light, which was, of course, unknown. The case of Iapetus was hardly reassuring in this respect, for half of this moon reflects ten times as much sunlight as the other!

Because both the masses and sizes of the satellites were uncertain, densities derived from them were even more so. About all that could be said was that the densities were lower than that of Earth, and probably even less than that of our Moon; in fact, they were not much greater than that of water ice. The idea of small "ice planets" gained support from observations of the efficiency with which these bodies reflected light at different wavelengths, particularly in the infrared. The reflection spectra resemble that of water ice, more or less dirty or clean as the case may be. At Saturn's distance from the Sun, the temperature of a body with no source of internal heat is so low that common water ice behaves like strong rock on Earth's surface. Thus, few expected much in the way of evidence for present or past geological activity on the satellites. However, Jupiter's Galilean satellites had produced some stunning surprises when viewed from close up, so who knew?

Titan was another story. Comparable to the planet Mercury in size, the only satellite then known to possess a substantial atmosphere, and of an unusual "orangish" color, this moon was the object of intense scientific interest and a prime target for the Voyagers. But as far as visual discoveries went, Titan was to prove a big disappointment.

In contrast, not much in the way of surprises was expected from the Saturnian ring system. Over the years, there had been reports of many minor divisions, weird dark "spokes," and other deviations from the apparently serene regularity of the classical A, B, and C Rings. Moreover, astronomers remembered all too well Percival Lowell's widely publicized "observations" of canals on Mars (and Venus as well!), features that turned out to be imaginary, and generally dismissed claims of detailed ring structure. As far as the actual thickness (thinness, actually) of the rings was concerned, the hope was that smaller limits could be set and that we could learn something about the physical sizes of the individual ring particles from their appearances at different aspects with respect to the Sun.

As mentioned earlier, a flyby probe is little help in discovering distant satellites of a planet. But might it reveal small, close-in moons, or even additional rings beyond the known rings? Telescopic observations from Earth provided some tantalizing hints. For several reasons, the best time to look for such things is when the rings appear edge-on as viewed from Earth. First, the prominent rings essentially disappear, so light reflected from them cannot overpower indications of a faint satellite or exterior ring. Second, one knows where to look, for any additional satellites or rings would almost certainly lie in the planet's equatorial plane, and thus would appear like beads on a needle, the needle being just where the main rings should be.

Such edge-on appearances occurred in 1966 and 1980, and in both cases there were claims of additional satellites and additional rings, the latter both inside the C Ring and outside the A Ring (the sequence moving outward from Saturn is C, B, A). Controversy raged over the reality of these objects, and though this is no place to go into the details, by 1979 or so there was general agreement that at least two additional rings existed.[6] One, called the D Ring, apparently filled the gap between the inner edge of the C Ring and the top of Saturn's atmosphere. The other, dubbed the E Ring, extended from outside the A ring to some 400,000 kilometers from the planet. Both of these features had very low surface brightnesses and thus, presumably, a very low density of particles.

Regarding the planet itself, just getting a closer look could not improve Saturn's generally bland appearance, but there was hope that there would be smaller atmospheric features—ones not visible from Earth—that could be used to determine motions in the upper atmosphere. For some time planetary astronomers had known that the equatorial regions of Saturn (or at least the regions of the upper atmosphere that we can observe from Earth) rotate much faster than the temperate regions. In this respect the ringed planet resembles Jupiter, except that in Saturn's case the velocity difference is much greater, amounting to a zonal wind of almost 1,000 miles per hour, and the expectation was that the Voyagers would fill in the details of this picture.[7]

There was justified confidence that scientists could learn a lot about the strength, orientation, and structure of Saturn's magnetic field, and that there would be a lot of hard information about the planet's radiation belts. These were not expected to be as intense as Jupiter's, for Saturn's rings would literally soak up all the energetic electrons and ions in the most concentrated regions of trapped radiation, and there was no known source of material such as Io's volcanoes, which spew roughly a ton of fine debris into space every second. There was also the hope (and it was just a hope) that closeup observations would detect radio signals that somehow were trapped in Saturn's magnetosphere and thus could not be observed from Earth, those that could not make it through our own planet's ionosphere, or ones that were just intrinsically weak. All in all, astronomers before the encounter knew only enough about Saturn and its system to make one wonder what the Voyagers might find.[8]

Just how the paucity of knowledge about the Saturnian system bedeviled those who actually had to plan missions is illustrated by differences that erupted in the 1970s over where to target probes (eventually to be Pioneer 11 and the Voyagers) for their closest approaches to the planet. For example, to get as close to Saturn as possible, there were proposals to attempt passages through the Cassini Division or inside the inner edge of the C Ring. Fortunately these were rejected, for these supposed "gaps" actually contain a substantial population of particles, and any spacecraft that tried to run them probably would be destroyed.

However, the more general question, How close can we get? (or, to put it another way, How far out are there rings that may demolish our probes?) was one that could not be dodged. This was a serious problem, and would not be settled definitively until a spacecraft flew by Saturn. Any such rings could not be densely populated, or else astronomers already would have detected them from Earth, but could they still pose a deadly threat? Engineers and mission planners needed hard numbers, but observers could give only educated guesses. Those estimates indicated that there was only a small chance that a probe passing outside of the A Ring would be destroyed, and they turned out to be correct (or, more accurately, "pretty much correct"), but there was more than a little bit of luck involved.[9]

The first spacecraft to reach Saturn was Pioneer 11, which made its closest approach on September 1, 1979. Of course, this was too late to be of any help to Voyagers' designers and builders, but the flyby provided much useful information for those planning exactly what to do in Saturn's vicinity. Perhaps the most important result was that the probe crossed the ring plane through the E Ring with no damage; things were looking up for Voyager.

Pioneer's crude imaging system failed to detect either the D or E Rings, and it showed little new detail on the planet and its satellites, but a new ring *was* discovered. Dubbed the F Ring, the inner edge of this very narrow (less than 800 kilometers) feature lies only 3,600 kilometers beyond the outer edge of the A Ring.

Images and the charged-particle detectors (which worked in this case by noting temporary decreases in particle abundances) provided evidence of one new moon (which may have been seen earlier from Earth) and possibly several more. Pioneer also found that Saturn emits about twice the energy it receives from the Sun, confirming earlier Earth-based observations, and that its magnetic axis lines up almost exactly with its axis of rotation, a dramatic difference from the cases of Earth and Jupiter, where the magnetic and geographic poles lie far apart. Pioneer also found that the rings have no more than three-millionths of Saturn's mass and are embedded in a tenuous cloud of hydrogen atoms, confirming indications from earlier Earth-based studies by radio and optical means that the ring particles are essentially water-ice cubes.[10]

Meanwhile, as the Voyagers cruised on toward Saturn, back at JPL the people who were actually going to operate the spacecraft during the encounters were practicing long and hard. They had to anticipate problems as best they could, for it would take radio signals over an hour to travel between Earth and the probe, and almost three hours for the round trip. It might prove to be too late to do anything effective in case of an emergency, but at least the time spent wondering what to do could be shortened as much as possible. During trial runs surprise anomalies were introduced at unexpected times, and simulations went so far as to see how the flight teams reacted to a major earthquake in southern California—a

Charles Kohlhase (*foreground*), James F. Blinn, and Patricia Cole create special effects for PBS's *Cosmos* series. (Courtesy of Charles Kohlhase)

very real possibility. Nothing that could be imagined was left to chance, but interestingly, and significantly, scientists and engineers encountered some things they could not have imagined.[11]

Ed Stone, Voyager's Project Scientist, invited Ellis D. Miner to become the Assistant Project Scientist for the Saturn encounters, a position that he eventually held for about twelve years. Miner had been born in 1937 in Los Angeles but grew up in the San Francisco Bay area and northern Utah. As happened in many cases, he first got interested in astronomy as a hobby and eventually decided to concentrate on astrophysics during his graduate study at Brigham Young University. After completing his doctorate in 1965, he fulfilled his two-year active-duty commitment in the U.S. Army. For that period he was assigned to JPL, reflecting the laboratory's past close associations with the Army missile program. Miner remained at JPL, serving as an experiment representative on the Mariner 6 and 7 Mars missions, and then continued to work with later missions, including the Mariner 9 Martian orbiter. He next worked with infrared experiments on the Viking Mars spacecraft.[12] Miner and Stone plunged into Saturn work much more quickly than anticipated.

The Saturn encounter had an unexpectedly early beginning. In January and February 1980, when Voyager 1 was still far from Saturn, it began to pick up radio bursts of very low frequency (very long wavelength) coming from the planet—emissions that cannot penetrate Earth's ionosphere. These bursts did not occur at random intervals but varied periodically in about ten hours and forty minutes. As in the case of a similar situation on Jupiter, that interval is the rotation period of the planet's magnetic field, and also of the inner, massive body of the planet, where that field is generated.[13]

As Voyager 1 neared Saturn, the spacecraft was essentially in the same operating condition as when it left Jupiter. One major exception was the photopolarimeter, which measured the brightness and the kind and amount of polarization of light. This instrument suffered a drastic loss of sensitivity before the Jupiter encounter, and was turned off before Saturn was reached.

In early October, long before Voyager 1's closest approach to Saturn (which occurred on November 12, 1980), images of the rings showed a completely unexpected number of "divisions" separating the main rings into numerous narrow (that is, of small radial extent) ringlets. As spatial resolution increased, the number of ringlets grew into the hundreds until the ring system looked like a phonograph record or a CD. As the narrowest ringlets were always just at the limit of resolution, whatever that was, it was probable that they numbered in the thousands. There are ringlets not only in the A, and especially the B, Rings, where visual observers had often glimpsed divisions, but also in the C and D Rings, which appear uniform when observed from Earth.

The relatively few divisions in the A Ring might just be explained by the gravitational effects of Saturnian moons, but the sheer number of ringlets in other regions, in particular the B Ring, suggests something more. A possible, if partial, explanation came from images of the well-known Cassini Division between the A and B Rings. This presumed "gap" actually contains a number of ringlets, and in its outer part is a series of alternating (in the radial direction) bright and dark features that can be explained as

Saturn's ring system taken from Voyager 1 on November 14, 1980. (NASA photo)

so-called spiral density waves, features attributed to the gravitational effect of the satellite Iapetus. Interestingly, this mechanism had originally been devised to explain the existence of spiral arms in galaxies such as our own Milky Way, but on scales of thousands of light-years instead of thousands of miles.[14] However, that process could not explain the vast majority of ringlets—what else was going on?

One possible explanation involved "shepherding" or "embedded" moonlets, small bodies only a few hundred or a few tens of miles across whose gravitational effects produce the observed structure. Unfortunately, with a few conspicuous exceptions, there was no evidence for such tiny satellites. In one case, a small satellite discovered by Voyager 1, 1980S28, orbits just outside the A Ring and may be responsible for that feature's sharp outer edge. Still, in sum, we have no generally accepted explanation for the complex structure of Saturn's rings.

Voyager demonstrated for the first time the complexity of this planetary puzzle. Yet there are more puzzles. For example, a division in the A Ring, called the Encke Division, holds a pair of ringlets that are not only discontinuous but also "kinky." How this could be is, frankly, unknown, and there was no trace of any small moonlets in the vicinity whose influence might completely account for such an appearance, although Voyager team

Wide-angle image of Saturn's rings taken by Voyager 2 on August 29, 1981. (NASA photo PIA-00534)

scientist Mark Showalter predicted the presence of the moonlet of Pan based on the narrow ringlets and later discovered it in Voyager images.[15]

The narrow F Ring presents a similar puzzle. Voyager 1 images showed that it was made up of several separate, very narrow, discontinuous strands that at one location appeared to be braided just like a pony tail! Voyager 1 did discover a pair of small "shepherding" satellites that can explain the F Ring's narrow extent, but they do not help much as regards its bizarre structure. Again, Showalter explained the F Ring appearance as a consequence of the "beat frequency" of satellites Prometheus and Pandora, which provide at least a partial answer.[16]

There are more general questions as well. Why, for example, does the B Ring have so many more subdivisions than the A Ring? Then, too, subtle color differences show that particles in the different rings have slightly different surface compositions. At the same time, some ringlets in one ring have the same color as the majority in another, indicating that material has not been "mixed" in a ring, and that some ringlets, at least, retain their identity over long times.[17] On the other hand, the very fine features appeared to change rather rapidly (just how rapidly may be established by the Cassini spacecraft, on its way to Saturn as this is being written).

A new and diffuse ring, dubbed "G," was confirmed by Voyager 1 images. This feature had revealed traces of its presence by way of a temporary decrease in the number of charged particles detected by Pioneer 11, but scientists were still pleased to actually see it.

The ring system looked startlingly different when viewed from different aspects with respect to the Sun. From Earth we can see it only in "back-scattered" light, where the illuminating rays behave much like an elastic ball bouncing back off of a wall; this is because our planet is always very much on the "Sunside" of Saturn. Things assumed a very different aspect when Voyager 1 looked "back" at the rings from outside Saturn's orbit around the Sun. Here the view was provided by "forward-scattered" light, the rays behaving like a flat stone skipping on water. To be specific, very small particles (like those in cigarette smoke; ordinary sand or dust grains are too large) scatter light quite efficiently in the forward direction, far more efficiently than larger ones (as drivers behind dusty windshields may have observed when heading directly into the sunset). This difference in behavior revealed that the E and F Rings are composed for the most part of very tiny particles, even though none of them could be observed directly. Using different techniques, radio measurements indicated that A Ring contains a substantial fraction of bodies in the size range of about 10 centimeters, the Cassini Division, 8 meters, and the C Ring, 2 centimeters. In all these cases it is important to note that in reality there is almost certainly a wide range of particle sizes involved; the numbers mentioned are only representative.[18]

When viewed from their unilluminated side (that is, the side facing more or less away from the Sun), the rings appeared strangely different. Regions densely filled with particles, such as the B Ring, were dark, whereas relatively (but not entirely) empty regions, such as the Cassini Division, appeared bright because sunlight filtered through them.

As to the thickness of the main rings, Voyager 1 could not give a definite answer. As with every method tried for centuries, the best that could be done was to estimate that they must be very thin. Indeed, there is the intriguing, but unfortunately improbable, possibility that these rings are only one particle thick. If so, an observer seated on a ring particle would have a novel view; rocks and pebbles in front, behind, to the left and to the right, but above and below only black space and the distant stars.

Completely unexpected was the appearance of dark, fuzzy radial "spokes" (like those of a bicycle wheel) on the B Ring. These are most prominent when they first emerge from the planet's shadow and gradually fade and become distorted as they orbit the planet in the sunlight. Interestingly, this distortion is exactly that expected from the decreasing orbital velocities at increasing distances from Saturn. Whatever the spokes are, they behave just like normal ring particles as far as their orbital motions are concerned. Amazingly, similar phenomena had been glimpsed as early as the nineteenth century by telescopic observers but dismissed as illusory.[19] The spokes appeared dark as Voyager approached Saturn, but on the way out, when viewed in back-scattered light, they appeared as bright features. They must, then, be composed of very small particles, but just what are they? Possibilities include some sort of "electrostatic levitation" (as, for example, when small pieces of tissue paper jump up to your finger after you have walked across a carpet on a very dry day), but as of now we just do not know.

The Titan encounter was looked forward to with much anticipation but turned out to be disappointing insofar as publicity value was concerned. However, there were interesting and perhaps valuable scientific findings. The satellite's all-encompassing clouds proved impenetrable to Voyager 1's gaze, and the resultant images merely showed a fuzzy orange ball with some subtle and unexciting features such as a darker north polar hood and several haze layers.[20] On the other hand, the way in which the probe's radio signals faded and then reappeared as it passed behind the moon as seen from Earth revealed that the atmospheric pressure at the surface was 1.6 times that on our planet. Because surface gravity on Titan is much weaker than on Earth, there is a lot more gas above every square inch of that satellite's surface than above us. In fact, the total mass of Titan's atmosphere is comparable to our own.

A number of different studies, such as the radio work just mentioned, along with infrared observations, showed that Titan's atmosphere was mostly molecular nitrogen, with perhaps a few percent of methane, along with whiffs of hydrogen and organic compounds, including those of nitrogen. It is not surprising that nitrogen should form the bulk of the atmosphere, for all the water and carbon dioxide are frozen out as surface rocks and most of the hydrogen and helium have escaped into space.

Titan's surface is a cold place, its temperature near the "triple point" of methane, where solid, liquid, and gaseous phases of that compound, at a certain pressure, can all coexist in equilibrium. Methane had been known to be there for decades, but now it appeared that this substance might play the same role on that moon as water plays on Earth; methane vapor, methane rain, methane seas, and methane ice might all be present. Moreover, ultraviolet light from the Sun will no doubt cause chemical reactions with the organic molecules known to be present, perhaps forming more complex, and heavier, hydrocarbon molecules that would then sink back to the surface and, in the process, explain the orange smog that envelopes Titan.

There are some exciting possibilities. Is Titan covered with a universal ocean of ethane and methane on which floats a thick layer of complex organic substances? In this "soup" are there clues as to how life began on Earth? Perhaps. Titan is so cold that no one expects any form of life to have developed there, but by the same token, that cold may have preserved some interesting evidence of early times that has not survived on our own planet. The idea of the universal ocean of ethane and methane, for example, has been proved wrong. In the near infrared, where the Voyagers had no imaging capability, we can see Titan's surface, and it is varied.

One must mention here that the nitrogen atmosphere, surface pressure, and surface temperature of Titan were all predicted with some accuracy by Donald M. Hunten of the University of Arizona prior to the encounter. This is one of the few times in the history of planetary astronomy that a prediction (other than those involving the motions of bodies) turned out to be precisely correct.[21]

Any disappointment there may have been about the lack of detailed features in images from Titan was allayed by the fact that this situation cleared the way for Voyager 2 to continue on to Uranus and, hopefully, Neptune. Had the second probe been targeted to obtain the best possible observations of Titan, that would have precluded a mission to those other planets. However, it was an easy decision to choose the chance of investigating two more planetary systems rather than investigating more pictures of a fuzzy orange ball.[22]

Astronomers had expected to obtain little information from the other satellites, but once again nature had some surprises in store. Between planet-sized Titan on the one extreme, and ring particles and isolated "flying mountains" on the other, Saturn has a number of companions of a size that previously had not been investigated close-up. Smaller than Jupiter's Galilean satellites and Earth's Moon, yet larger than Mars' attendants Deimos and Phobos, these bodies were really terrae incognitae.

Astronomers were certain that temperatures had never been very warm this far out in the Solar System, so that even relatively small bodies could have retained light molecules such as H_2O, though of course they would have lost most of their helium and elemental hydrogen. Most of them appeared to keep the same face toward their parent planet, as does our own Moon, and a variety of studies (estimates of density, reflectivity at various wavelengths, and so on) indicated that common water ice was a major bulk constituent, and perhaps the overwhelming constituent in some cases.

Iapetus was known to be weird; the trailing hemisphere as it orbits Saturn is as bright as the brightest snow, and the leading hemisphere was as dark as the darkest coal. The common assumption was that some black material was covering part of an otherwise white ice surface; the question was, and is, does that material come from inside the moon or somewhere outside?

Alas, Voyager 1 could not answer that question. Images showed nothing on the leading hemisphere, which looked as if it had been "airbrushed" by running into something like a cloud of black bugs on an auto windshield. However, the dividing line between bright and dark was sometimes irregular yet sharp, as if material had welled up from the interior and filled or partially filled craters or maria. At a loss, scientists could only hope that better images from Voyager 2 might resolve the paradox.

Hyperion yielded only low-resolution images (just some seven picture elements across), and it was not possible to determine its rotation period. The other substantial moons in order outward from Saturn include Mimas, Enceladus, Tethys, Dione, and Rhea. The surface of Mimas, the closest "large" moon of Saturn, is dominated by a huge crater, evidence of an impact that almost destroyed the satellite. In some images Mimas bore an eerie resemblance to the "Death Star" that appeared in the original "Star Wars" movie. There were many smaller craters as well, but their size distribution differs in different regions, for reasons that are currently uncertain. These craters are, in general, much deeper than those on our Moon or Jupiter's icy (that is, not including Io) Galilean satellites, probably because Mimas is so small (only some 400 kilometers in diameter) that its surface gravitational field is weak and water ice "creeps" but little.

For the most part, the surface of Mimas is saturated with craters, so that any new one would, on the average, destroy an older one. This situation implies that most of the surface must be relatively old (say, three or four billion years), though just how old is unknown in the absence of rock samples that can be dated by physical methods. One exception is the "monster" crater, later dubbed Herschel, which has not been modified extensively by other impacts and thus must be younger than most of the moon's surface.

In addition to craters, Mimas shows grooves, some of which are almost 100 kilometers long, roughly 10 kilometers wide, and 1 or 2 kilometers deep. Possibly these grooves were formed by the impact that created Herschel, or perhaps they are the result of crustal motions during an early time when the moon was still warm and its crust may have been mobile, but nothing is certain. On the whole, Mimas has a density just a bit (1.2 or so times) that of water, and its brightness is consistent with a surface covered with common ice.

The next substantial satellite outward, Enceladus, is a little larger, some 500 kilometers across, has a density 1.1 times that of water (or, effectively, water ice), and has a high reflectivity. The best Voyager 1 images were taken from a distance of 622,000 kilometers, and at this range no craters were visible, though a few enigmatic bright streaks showed.

Next outward is Tethys, a larger moon with a diameter of over 1,000 kilometers, a density equal to that of water, and a reflectivity roughly that of snow. Apparently it is almost pure water ice. In this case as well, the best Voyager 1 images were only of relatively low resolution, and aside from indications of some variation in surface brightness and strong hints of craters, the only prominent feature seen was an enormous valley that extended across a good portion of the Moon. Once again, this tantalizing hint awaited better images from Voyager 2.

Dione is almost Tethy's twin in size, though it is a bit larger and has a higher density of 1.4 times that of water. Its surface, on the average, is as bright as snow. But Voyager 1 images revealed that there were large regions that varied greatly in surface brightness, with a "front-to-back" difference in reflectivity exceeded only by Iapetus. However, on Dione, it was the trailing hemisphere that was darker and the leading hemisphere brighter, just the opposite of the situation on Iapetus.

The dark hemisphere of Dione revealed relatively few craters, but did display long bright "wisps" that in some cases were shown to be composed of a number of narrower bright rays. However, unlike the bright ray systems connected with lunar craters such Tycho and Copernicus, which are sprays of impact debris that clearly radiate from a common center, these features appear to be deposits of common ice. Perhaps they resulted from eruptions of liquid or gaseous water along vast rifts that may have opened at an early date when Dione's crust was mobile. On the other hand, such cracks may have been opened by a large impact. In any case, these bright wisps divide the darker portion of Dione's surface into irregular polygons that remind one of Jupiter's satellite Ganymede or, on a much smaller scale, ice floes on Earth.

Dione's bright side is *not* saturated with impact craters, so some process must have resurfaced it at some time. In fact, here there are two types of terrain, one much less densely cratered than the other, clearly the former is younger in relative terms, and must have been resurfaced, possibly by eruptions of water "lava" from below the surface. In addition, both polar regions of this satellite harbor long, wide, and deep valleys.

Rhea, with a diameter of over 1,500 kilometers, is second in size only to Titan among Saturn's satellites. Voyager came within 59,000 kilometers of this moon, providing the highest resolution views of any major satellite. Its density is 1.3 times that of water and it, too, has an overall reflectivity similar to that of new-fallen snow. The side of Rhea facing Saturn is heavily cratered, whereas the other side is streaked with lighter-colored (with respect to their background) wispy streaks.

The leading hemisphere in Rhea's orbital motion is relatively bright and for the most part uniform and bland. The trailing hemisphere is darker, as in the case of Dione (and, again, the opposite of the situation on Iapetus) and, again as in the case of Dione, laced with bright, wispy streaks. The bright side is, in general, heavily cratered, especially in the equatorial region, and the walls of some craters show patches of bright, white material, probably ice newly exposed by some process.

The images with the highest resolution were those of Rhea's north polar regions, where bright and dark regions are separated by a relatively sharp though irregular boundary that does not lie along the line dividing the leading and trailing hemispheres. Here large craters are much less abundant in the dark terrain, strongly suggesting that some process has removed them and that these regions are relatively younger than the brighter ones. Rhea's equatorial regions provided their share of puzzles such as smooth plains relatively free of craters and areas of slim, linear grooves.

Taken as a whole, the Voyager 1 observations revealed that, in general, Saturn's major satellites have had extended and complicated histories of surface formation, a completely unexpected finding. Some of these differences may be due to different internal compositions, some may be the result of tidal forces exerted by other satellites, and some to different distances from Saturn. Then, too, there is always the possibility that some unlikely event such as impacts of myriad fragments from a disrupted comet nucleus may have greatly altered at least part of a particular moon's surface.

Voyager 1 also helped to sort out the confusion regarding six relatively close Saturnian moons. We have already mentioned the F Ring shepherds and the A Ring outer shepherd, discovered by Voyager 1. In addition, this probe confirmed the "co-orbital" satellites, first observed from Earth, that lie between the A Ring and Mimas and share almost identical paths. In fact, their orbits are so similar that collision seems inevitable, but detailed calculations indicate that they never do come in contact. Voyager 1 confirmed that another small object, also found from Earth and named Helene, shares an orbit with Dione, leading it by some sixty degrees. Based on telescopic observations, there were numerous other possible satellites, but the above were the ones seen by Voyager 1. All were irregular rocks in shape.

As far as Saturn itself was concerned, the view was much blander than at Jupiter, evidently due both to the planet's thicker haze layer and the cloud tops having intrinsically less contrast. As on Jupiter, there are alternating bands of light and dark clouds, but this structure extends much farther from the equator. In contrast to Jupiter, however, Saturn's winds do not appear to correlate with band structure. The strong equatorial wind agrees well with the usually bright belt centered on the equator, but velocities generally drop off smoothly with increasing latitude. Specific cloud features were few compared to those on Jupiter, but there were some, such as "white ovals," which were used to determine rotation rates, and from them wind speeds with respect to the body of the planet.

As was expected, aurorae were observed in Saturn's polar regions. On the other hand, there were no images of Saturnian lightning, for the rings reflect enough light onto the dark side of the planet to drown out such discharges. However, there was evidence from several sources of enormous electrical discharges, possibly from the rings. Such "bolts" would be invisible because the rings do not have the atmosphere that alone could make such a phenomenon put on a luminous display. However, most of the Voyager scientists ultimately attributed the discharges to activity in the super-rotating equatorial atmosphere of Saturn.

Once Voyager 1 finished its observations of Saturn and its system, it headed through the outer reaches of the Solar System toward the depths of interstellar space. The instruments on the moveable scan platform were turned off on December 19, 1980, to conserve electrical power produced by the steadily deteriorating RTGs. The equipment that was shut down included imaging devices that were of little use, for there were no known celestial bodies along the flight path. Left running were experiments that might detect the expected transition from the solar wind to true interstellar space, for the spacecraft's mission had not, and has not, ended.

But even as Voyager 1 disappeared from the glare of publicity, its findings at Saturn caused a major change in plans and expectations for Voyager 2's encounter. Fortunately there was time to make such alteration, for the later probe would not make its closest approach until August 26, 1981, more than nine months after its predecessor, although, as might be expected, that was not time enough time for everyone involved.

Among the objects of increased interest the rings held first place, for their incredibly complex structure was completely unexpected, essentially unexplained, and altogether fascinating to scientists and the general public alike. Voyager 2 held great promise in this respect for its photopolarimeter was working, though only partially. Scientists hoped to use this instrument to observe the brightenings and darkenings of the star Delta Scorpii as it passed behind Saturn's rings as seen from the spacecraft; the technique held out the possibility of attaining a resolution of a mere few hundred meters.

Voyager 1 revealed that the major satellites are much more complex and interesting objects than anyone had previously expected, and scientists aimed to get different (and, in some cases, better) views of these bodies. There also would be attempts to confirm other small reported moons, to search for new ones, and to try to find any moonlets that might explain ring structure. Voyager 2 also carried an imaging system that was more sensitive than the one on its predecessor, and there was considerable anticipation that, this time, there would be more detail visible on Saturn itself.

On the down side, Titan became an object of less interest. Although it is one of the most interesting, and possibly significant, bodies in the Solar System, it was clear that Voyager 2 could not provide any significant new knowledge about the satellite. So, as we have seen, optimal coverage of Titan was abandoned in lieu of the opportunity to go on to Uranus and Neptune.[23]

One of the most spectacular results of the entire Voyager mission was the incredibly detailed resolution of Saturn's rings provided by photopolarimeter observations of their occultation of Delta Scorpii, though by this time planetary scientists were ready to accept almost any degree of substructure in the rings. Where images from Voyager 1 had shown a single ringlet, Voyager 2 often revealed myriad, even narrower ringlets; the numbers ran into the thousands. (In a related matter, we should mention that Voyager 2, like Voyager 1, turned up no evidence of any "embedded" satellites that might help explain such structure.) Subsequent analysis of the Voyager 1 radio science ring occultation eventually provided comparable or better resolution than the Voyager 2 polarimeter measurements (although at a much different wavelength). Just how long these minute divisions last is uncertain, for there was only the one opportunity to view a stellar occultation. The occultation observations provided positive confirmation that spiral density waves excited by Saturnian satellites cause some of the ring structure, but needless to say, there was, and is, no ready explanation for the majority of the ringlets.[24]

The photopolarimeter results also provided yet another upper limit to the thickness (perpendicular to the ring plane) of at least some features in the rings. The dimmings and brightenings of Delta Scorpii as it was occulted by various ring features were abrupt, in some cases just the response time of the instrument. These results implied that the ring thickness was less than 200 meters (600 feet).

Voyager 2's more sensitive imaging system did indeed reveal more atmospheric features than did that aboard Voyager 1. In these improved views, the ringed planet more closely resembled Jupiter. The "white ovals" similar to those on Jupiter proved to be long lived and anticyclonic, that is, high-pressure areas. There were alternating belts of winds blowing in opposite direction at increasing latitudes, though this pattern occurs at higher latitudes on Saturn. The fact that high-altitude winds on Saturn are predominantly eastward suggests that these currents extend downward 2,000 kilometers, and perhaps much deeper. Voyager 2 also imaged features that resemble circular storms on Earth, as well as rising convective clouds similar to those seen on Jupiter and in terrestrial cumulo-nimbus clouds.

Voyager 2 imaged all seventeen satellites of Saturn that were known at the time (as well as an eighteenth satellite later identified as Pan), though, of course, with various degrees of spatial resolution. The spacecraft took the first informative images of the outermost moon, Phoebe, which orbits the planet in a retrograde direction (opposite to the rotation of Saturn, and to the motions of ring particles) and is generally believed to be a captured asteroid. This moon turned out to be roughly spherical with a diameter of some 220 kilometers and rotating with a period of nine hours. Thus it is not tidally locked to Saturn and so does not always keep the same face toward the planet, a fact not unexpected because of the satellite's great distance from its primary. Phoebe's surface is about as dark as coal and probably dark red, properties consistent with a class of asteroids believed to be relatively unmodified since their formation. The surface has regions with significant brightness variations, but the images were not good enough to reveal individual features such as craters.

Better images of Iapetus did not resolve the quandary revealed by Voyager 1: did the dark material come from outside or inside the satellite?

Images of Hyperion revealed a very irregularly shaped object about 410 by 260 by 220 kilometers that had some "sharp edges" and other angular features. Its appearance suggests that the present satellite is the remnant of a once-larger body that has lost substantial chunks through collisions. This moon is one of the darkest objects in Saturn's system, and in fact the darkest that also shows spectral features due to water ice. The dark material may come from Phoebe, a source that also has been invoked to explain a possibly similar substance on Iapetus. If so, and if Hyperion is not tidally locked to Saturn, it should not have two distinct hemispheres, which it apparently does not. However, its actual rotation period remained unknown even after Voyager 2, due to its weird shape, chaotic tumbling, and the relatively short interval during which it was observed.

Voyager 2 never got very close to Mimas, Dione, and Rhea and added little new information to our knowledge of those satellites. The situation was more favorable in the case of Titan, but as in the case of Voyager 1, that planet-sized moon remained hidden under its haze and clouds, exhibiting only subtle atmospheric features.

Images of Tethys revealed some dramatic new features. There were several large impact craters and a single enormous one, Odysseus, which is between 400 and 500 kilometers across, almost half the diameter of the satellite itself and larger than Mimas! Despite its size, this crater has very little vertical relief; evidently solid "creep" in the ice-rich satellite has flattened out the topography. Ithaca Chasma, the huge valley or rift discovered by Voyager 1, proved to extend at least three-quarters of the way around the satellite. In addition, different areas of the surface have different crater populations, and, in particular, there are flat plains with relatively few craters, plains that may be evidence of floods of water at some earlier time. There was also evidence of more complex terrain mingling impact craters, hills, ridges, and valleys.

Voyager 1 had shown little surface detail on Enceladus, and guesses as to what better images might reveal ranged from little or nothing to very unusual things indeed. In the event, Voyager 2 revealed that this small object had the most active and complex geologic history of any of Saturn's satellites (always excepting what Titan might turn out to be like). There are many different types of features on various parts of the surface, ranging from heavily cratered areas to ones that are essentially crater free and thus must be very young indeed on the scale of geological time. Indeed, Enceladus may still be active geologically.

Some parts of the crater-poor plains are crossed by sinuous ridges and valleys that suggest similar features on Jupiter's satellite Ganymede. In both cases the favored interpretation is that this structure is the result of moving plates or slabs of ice floating on a liquid layer below. If Enceladus has been geologically active recently, or even now, the obvious question is, Where does the energy in the form of heat come from at this late date? The satellite is so small that any initial heat of formation must long ago have radiated out into space. Then too, this moon's density is so low that it can never have had a copious supply of radioactive elements. One attractive possibility is tidal heating forced by Dione, similar to the mechanism that makes Io such an active body.[25]

In sum, the Voyager 2 results strongly reinforced those of Voyager 1; namely, Saturn's major satellites have had complex histories that vary from body to body. Since the flybys, there have been a large number of studies of different satellites dealing with facts such as the relative number of crater counts, different bombardment episodes, multiple episodes of resurfacing, heating mechanisms, and the like. From this work have come some interesting scenarios concerning what has gone on in this region of the Solar System over the past four billion years or so. However, all these results are tentative, and none can supply any definite dates. As of this writing scientists have such dates only for terrestrial rocks, meteorites, lunar samples, and a few pieces of Mars. Except for results concerning our own planet, these dates are few, and they all relate to the inner Solar System, within Jupiter's orbit, and there is no guarantee that things happened the same way farther out.

Voyager 2 also imaged the eight small satellites (as opposed to the larger, "classical" ones known for almost a century) known or suspected at the time of the flyby. The A Ring shepherd and the two F Ring shepherds had been discovered by Voyager 1. Telescopic observations from Earth had revealed the others: the co-orbital satellites mentioned above, the "Lagrangian" in Dione's orbit, and the leading and trailing Lagrangians that share an orbit with Tethys. All are more or less irregular in shape, suggesting that they have been battered by collisions over the aeons, and they appear to have a range of reflectivities suggestive of different surface compositions.

Some twenty minutes after Voyager 2 made its closest approach to Saturn, it made its closest approach to Enceladus. Just after that event three things happened within a short time: the probe's trajectory caused it to disappear behind the planet as seen from Earth, cutting off direct radio contact; the spacecraft passed into Saturn's shadow, which would rapidly lower most temperatures aboard the spacecraft (and later raise them when the shadow had been traversed); and the plane of the rings was crossed. None of these events seemed to be cause for great worry, as similar events had not caused any catastrophes to previous spacecraft. Still, there was some apprehension.

A minute or two before midnight on August 25–26, 1981, Voyager 2's radio signal began to come in—the probe had survived! But jubilation soon turned into emotions entirely different, for something had gone seriously wrong. Telemetry indicated anomalous readings from several instruments, and the spacecraft's attitude-control thrusters had fired several times when they were not commanded or expected to. However, the most chilling discovery was that the scan platform, which held and pointed the imaging cameras, photopolarimeters, and spectrometers for both the ultraviolet and the infrared, was frozen in position. More precisely, it could still move in elevation (up and down) but not in azimuth (side to side). After a few hours other problems had cured themselves or gone away in some manner, but the scan platform was still crippled.

What saved the situation was that the platform could still be moved in azimuth if that motion proceeded at slow speed. Evidently something had literally "gummed up the works" and affected the driving gears. Candidates were lubricant that had somehow migrated from where it was supposed to be and had to be "migrated" back some way or some foreign object, such as a teflon screw or some styrofoam insulation, that had somehow been dislodged from its proper location and had to be slowly ground up by the gears involved. In addition, around the time of ring crossing, the plasma experiment detected what appeared to be the result of myriad tiny particles vaporizing on impact with the spacecraft. Then too, the unexpected thruster firings suggested that particles just a bit larger had hit the probe and knocked it askew. Ellis Miner suspects that the gear froze to the shaft as a result of the loss of lubricant due to too frequent use at high rates. Eventually the shaft, which was press-fitted into the housing, broke free, allowing lubricant to migrate into the fitting and serving as the new slip surface. What actually happened probably will never be known.[26]

But whatever did happen, mission controllers were able to work around the difficulty and get the scan platform back to work, though admittedly slowly. Scientists lost some high-resolution images of Enceladus and Tethys, along with good views of Saturn's southern hemisphere, but things came under control sufficiently to provide images of Phoebe some nine days after ring passage.

For the Voyagers, there were many miles to go, and many things yet to do. Despite the problems during the Saturn encounters, it seemed that Voyager 2 might actually make it to Uranus in reasonable working order. But that journey would take four and a half years, and much could happen in that time.

11

URANUS, NEPTUNE, AND BEYOND

Voyager 2 now coasted toward Uranus, heading ever deeper into unexplored interplanetary space. Little was known about the outermost planets, and less about the space environment through which the spacecraft navigated; new hazards and unanticipated difficulties confronted Voyager mission operators located on far distant Earth. As had been true since launch, the two exploring spacecraft and their operating, engineering, and science systems were a "work in progress." Though contingency planning, remote programming, innovation, and ingenuity gave the probes greater operational, engineering, and science capabilities than they possessed when they left Earth in 1977, Uranus and Neptune and their space environments created unforeseen challenges.

Despite two centuries of effort, astronomers had learned little about Uranus, although it was the first planet to be discovered in the modern era. Mercury, Venus, Mars, Jupiter, and Saturn are bright, naked-eye objects and have been known since humans first gazed intelligently at the sky, but the greenish giant Uranus had escaped notice, even though it is just visible to a keen, unaided eye in a clear, dark sky.[1]

William Herschel discovered Uranus, which he first called the Georgian Planet (named after George III of Great Britain) on March 13, 1781. During the nineteenth century Uranus attracted a great deal of attention, but most of the interest had to do with that body's orbital motion around the Sun. The planet's seeming departure from the laws of celestial mechanics eventually resulted in the spectacular prediction and discovery of Neptune, but there were nonetheless few advances in our knowledge of physical conditions on and near Uranus itself.[2]

At the time the Voyager spacecraft were launched, five satellites of Uranus were known to exist. Their motions revealed that the planet was a giant similar in many respects to Jupiter and Saturn, though with a substantially greater density, implying a smaller fraction of hydrogen in its makeup compared to the two larger planets. The green planet's axis of rotation was known to be "lying on its side," but the rotation period itself was uncertain, as no definite atmospheric features had ever been observed, although spectroscopic methods suggested something between sixteen and seventeen hours. There were sporadic claims of planetary rings (the first by Herschel), but these were generally assumed to be spurious. There was no evidence for any Uranian magnetic field or radiation belts, though both were suspected to exist. In a similar vein, impulsive radio emission that might be associated with such features had not been observed, so there was no hint as to the rotation rate of the inner body of the planet. Molecular hydrogen and methane were known to exist in the atmosphere, and selective absorption due to the latter substance gave the planet its distinctive color. It was generally believed that one or more cloud decks composed of ice crystals of some sort (methane, perhaps) floated somewhere down in the outer gaseous envelope.[3]

The satellites of Uranus, of course, were mere points of light, and values of their sizes, masses, densities, compositions, and reflectivities were only rough estimates. There were indications that their surfaces were covered with ice and snow, and that their densities were just a little greater than those of water, similar to those of Saturn's icy moons. The diameters of the Uranian satellites were estimated to lie in the range from a few hundred to 1,000 kilometers (similar to the larger asteroids), but their compositions, surface appearances, past histories, and just about everything else about them were complete blanks.[4] In sum, what we knew about Uranus before Voyager 2's flyby had not profited a great deal from the remarkable rebirth of planetary astronomy that was fueled by funds from NASA starting in the early 1960s. There was however, one spectacular exception: the discovery of the planet's rings in 1977.[5]

At that time, no one was looking for rings around Uranus, or even considering such a possibility, inasmuch as Saturn was then the only planet known to have such appendages. What observers *were* seeking was information about the physical state of Uranus's upper atmosphere, and in particular some clue as to the amount of helium present. The technique used measured the rate at which light from a distant background star faded out and then reappeared as the planet moved over it as seen from Earth. Such "occultation" measurements by a number of observing teams indeed generated the information sought, but investigators also found a number of sharp, definite, "narrow" (they did not last long) brightness dips both before and after the expected central one. The rings had been revealed!

Data from the initial occultation and later events of the same kind provided a picture of a bizarre system. There were at least nine rings about Uranus and all were incredibly narrow, some only a few kilometers in radial extent and only one as much as 100 kilometers wide. Moreover, some of these features were not circular but elliptical. Many were inclined at small but different angles to the planet's equator and varied in width around the planet. The widest, outermost ring, dubbed ε (epsilon), produced most of the reflected light from entire ring system. To top it off, the ring particles were very dark, reflecting only some 2 percent or so of sunlight, in vivid contrast to the bright, ice-covered moonlets that orbit Saturn.[6]

As well as voyaging into the realm of the outermost planets, Voyager 2 entered the unknown in another sense, for the environment of the remote Solar System changed markedly from that of the nearer planets. It was colder and darker; at Uranus sunlight is four times weaker than at Saturn and almost four hundred times weaker than at Earth. As a result, to take one example, exposure times for photographic images must be correspondingly longer than is the case closer to the Sun, a problem that is especially serious in the case of the Uranian rings, which are intrinsically darker than coal. In turn, these longer exposure times increase the "smearing" of an image due to the motion of the spacecraft, and although this degradation can be reduced by causing the imaging system to actively "track the target," such a solution involves slowly rotating the craft in just the right direction and at just the right speed, a tricky maneuver at best. In addition, there was some worry that the scan platform could seize up again. Moreover, because the scan platform employed a stepping motor, the incremental movements at low speeds needed for tracking could not be achieved. Thus, mission managers decided it was not possible to use the scan platform.[7] It should be emphasized that just getting closer to Uranus does not increase its surface brightness or shorten exposure times; the planet just gets bigger in apparent size.

Weaker sunlight also means lower temperatures, and although critical spacecraft components were heated by electrical power supplied from the RTGs, the demands on these systems were clearly much greater. Moreover, colder temperatures required that heaters be left on longer, and communications and other systems required greater power supplies at greater distances from Earth. The possibility of mechanical failures also grew geometrically as the space environment became more hostile. Although it was reassuring that Voyager was then one of four ships operating at similar distances from the Sun, Pioneer 10 and 11 were much simpler machines, and Voyager 1 was essentially "buttoned down" in cruise mode. There was very real concern about how well Voyager 2 would perform at Uranus.

Increased distance from the Sun also meant that radio signals from Voyager 2 would be weaker when received on Earth. All other things remaining the same, this reduction would necessarily reduce the rate at which data could be transmitted, a decrease that presented a particularly serious impediment for images, each of which requires a relatively enormous number of data bits compared to that required for the transmission of engineering or other scientific data.

Some things could be and were done on Earth to improve communications. For example, antennas at ground-tracking stations were "ganged together" in arrays to provide what was in fact a much larger collecting area than any individual dish could provide. On the spacecraft end, several effective measures were taken, although obviously the probe's primary antenna could not be enlarged, nor could its transmitter power be increased. The improvements relied on the fact that most modes of data transmission (alas, usually the simplest ones) contain a great deal of "excess" information (in the sense that it is not actually needed; for example, black sky outside a body's limb); the trick is to somehow transmit only the essential information. This was done in two ways. First, formatted data was coded to reduce the possibility of bit errors during transmission using a Reed-Solomon encoder, a piece of hardware that was launched aboard the Voyagers but had not been used before, serving only as a backup. It did not function at the high transmission rates used for data return during the Jupiter and Saturn encounters (i.e., >50 kbps). Second, an on-board data compression system using the backup computer of the flight data system reduced the data rates needed to transmit real-time images by 70 percent.[8] There were also improvements in operating the attitude-control system, so that the angular rates at which the spacecraft made small corrections to maintain its orientation were reduced by a factor of two or three, which resulted in less smearing of images compared to those taken of Jupiter and Saturn.[9]

Because of the unusual orientation of Uranus's axis of rotation, the planet's South Pole was pointed toward the Sun during the latter part of the twentieth century. As a result, Voyager 2 would view basically only the planet's illuminated Southern Hemisphere as it approached, and the dark Northern Hemisphere on the way out, and scientists expected to find substantial atmospheric temperature and perhaps structural differences between the two regions. Similarly, the ring and satellite systems would be flown through from "bottom to top" (or south to north).

The primary consideration in choosing Voyager 2's flyby path was the need to deflect it in such a way as to send it on to Neptune after the Uranus encounter. However, there was some timing flexibility within that overriding target requirement, and this leeway was used to time the flyby to provide a close pass to Miranda, as well as planetary and ring occultations at the best times for data reception back at Earth (because Uranus was then south of Earth's celestial equator, it was important to use the tracking station at Canberra, Australia, for the occultations). The stage was set for the Uranus encounter, the closest approach to the planet's center of 107,000 kilometers planned for January 24, 1986. Voyager 2 began its encounter activities on the preceding November 4, when it was still some 10.3 million kilometers away.[10]

Miranda, taken by Voyager 2 on January 25, 1986. (NASA photo PIA-01490)

Although the Uranus system certainly provided its share of surprises, the appearance of the planet itself on direct images was not one of them. As expected, there was little atmospheric detail, the general impression being that of a bland, greenish billiard ball. However, there were a few wide, diffusely edged, brighter and darker bands centered on the South Pole. In addition, there were a few bright, discrete features suitable for determining rotation periods. Evidently these structures were high-level clouds of methane ice that topped upwelling convective plumes, similar to the anvil heads that cap cumulo-nimbus thunder heads on Earth but on a vaster scale. The derived rotation periods were just under 17 hours at latitude about twenty-seven degrees south to 14.2 hours at south latitude seventy degrees. The decrease in rotation period in the equator-to-pole direction appeared to be smooth, but the small number of features observed made that conclusion only probable.

If Uranus images produced results that were mundane by comparison with those of Jupiter and Saturn, other techniques more than made up the difference. Some five days before closest approach, Voyager 2 detected bursts of low frequency (in the kilohertz or kilocycles/second range) radio emissions. There were a variety of different types of emission at different frequencies, no doubt produced by different mechanisms at different locations in the Uranian system. One component varied with a period of about 17.24 hours, which is almost certainly the rotation period of the planet's magnetic field, and thus of the interior of that body.

Measurements of Uranus's magnetic field were a revelation. As was expected, the planet has a magnetic field, but the magnetic poles are offset some sixty degrees from the poles of rotation. The situation is even more complex than that: if the field is thought of as arising from a simple bar magnet (of course it does not, but this is a convenient approximation), not only is the long axis of that bar inclined with respect to the planet's axis of rotation, but the entire bar is offset from the center of Uranus by some 30 percent of that body's radius.[11] This unusual geometry produces some interesting effects. For one, even though Uranus's magnetic field is about as strong as Earth's, a balloon-borne observer riding in the planet's atmosphere would measure complex and dramatic variations in field intensity from place to place. For another, the rings and satellites will absorb particles in different regions of Uranus's radiation belts in complex manners that vary during the day and during the year.

Infrared spectra and radio occultation (the fading and subsequent return of radio signals from the probe as Voyager 2 passed behind the planet as seen from Earth) combined to give more information. Surprisingly, the temperature at the cloud tops above the South Pole (which had been in continuous sunlight for decades) was then about the same as that at the equator and the North Pole (which had been in unrelieved darkness for decades); evidently some form of mass motion transfers solar energy from one place in the atmosphere to the other. In addition, helium constitutes about 15 percent (by number of molecules) of the upper atmosphere, about the same fraction as in the Sun, and there is evidence of a cloud deck of methane-ice crystals down where the pressure is about that on Earth's surface.

Uranus appears to have no substantial source of internal heat. This contrasts with Jupiter, which still retains a great deal of its initial heat of formation, and Saturn, where the energy released by the sinking of heavier helium through the mostly hydrogen outer layers may provide a substantial heat source. Further, long-exposure images of the dark hemisphere of Uranus showed no evidence of lightning, which was not unexpected given the low level of visible convective activity. No aurorae appeared on these direct images, but ultraviolet observations revealed an auroral "cap" centered on the South Magnetic Pole.

Everyone, including scientists, engineers, reporters, and the general public, looked forward to what might be revealed about the rings of Uranus from the images and optical and radio occultations.[12] Based on what the Voyagers had revealed about Saturn's rings, scientists anticipated some exciting new findings—and they were not disappointed.

Voyager confirmed the existence of all nine rings that had been observed from Earth and added two more. One was a faint and narrow (only 1 or 2 kilometers wide!) feature just inside the ε Ring; the other was a relatively wide (some 2,500 kilometers), extremely tenuous annulus whose outer edge lies just inside the innermost ring discovered from Earth. The rings themselves appear to be composed of bodies in the size range from, say, soccer balls to single-family houses, with no evidence for any concentration of tiny dust particles. However, images taken after closest approach, in forward-scattered light (Sun behind the rings as seen from Voyager 2), revealed an extremely tenuous disk of dust particles pervading the entire ring system, a disk that contains a surprising amount of brightness variation in the radial direction.[13] Whether these dust structures should be called "rings" is a moot point.

What is not debatable is the noticeable lack of dust concentration in the rings, whose relatively substantial particles are the only obvious source of the dust. A source there must be, however, and one that is producing dust today, for a variety of mechanisms, such as drag due to Uranus's wraith-like outer atmosphere, will cause tiny dust particles to spiral down into the planet in a few thousand years at most (the relatively huge ring particles would not be so affected.)

Ring occultations of the stars Sigma Sagitarii and Beta Persei as observed from the spacecraft revealed a confusing variety of ring structures (or substructures) that might be due to partial rings or "arcs"; the data were not sufficient to determine just what these structures are. However, a number of the rings showed small-scale fine structure reminiscent of Saturn's B Ring, and the ε Ring in particular revealed features strongly resembling those expected from spiral density waves. Some of the rings have extremely sharp edges, in the range of a few tens to a few hundreds of meters, and the sharp outer edge of ε, in particular, suggests that this feature is no more than 150 meters thick.

The nature of the very dark surfaces of the ring particles is still a puzzle. One suggestion is that they are coated with methane ice that has been darkened by bombardment with energetic charged particles in Uranus's radiation belts, but the true explanation is presently uncertain.

Before the Uranus encounter, scientists anticipated (or had a strong hunch) that Voyager might discover at least 18 new, small satellites, with one shepherding moon orbiting just inside, and another just outside, each of the then nine known rings. Ten new, tiny moons were indeed discovered, all with very dark surfaces, but for the most part they did not orbit where expected.

Two satellites shepherd the ε Ring, the outer one being the largest of new moons at some 170 kilometers across. Voyager 2 took a medium-resolution image of this body and found it to have a uniformly dark, crater-pocked surface; surprisingly, it appeared spherical, unlike the irregular shapes of other Solar System bodies in this size range. The inner shepherd was tiny, no more than 60 kilometers across. Thus, in this case the shepherding explanation for a narrow ring, as well as the density wave nature of at least some of the ring substructure, was confirmed.

No other shepherds were found (although some have been suspected) despite a concerted search; in fact no other satellites were found anywhere in the ring system. Such moons may indeed exist, but they would have to be less than about 10 kilometers in diameter to have escaped detection. Thus there was a loophole left for the shepherding theory, but only a little evidence for the mechanism in action as regards Uranus's ring system.

The other eight satellites discovered by Voyager 2 all move in Uranus's equatorial plane between the outer edge of the ε Ring and the orbit of Miranda, the innermost of the moons known before Voyager 2. Without exception they are small, less than 100 kilometers across, and have dark surfaces (in contrast, Saturn's small moons all have bright surfaces). Spacecraft images resolved none of these satellites, so estimates of their sizes and reflectivities are just that—estimates.[14]

In contrast, the five largest moons of Uranus revealed a great amount of detail. All of these satellites are smaller than Earth's Moon, so on the one hand one might expect little in the way of geologic activity on these bodies. On the other hand, the "icy" satellites of Saturn that have similar sizes exhibit a wide variety of geologic activity that appears to have extended to relatively recent times in some cases. Earth-based observations indicated that the surfaces of these bodies were composed of mixtures in various proportions of water ice and some dark material of unknown composition but gave no hints as to what those surfaces look like.

Voyager observations provided the first good estimates of these moons' diameters, masses, and densities. All have densities from 25 to 60 percent denser than water (or water ice), and appear to be composed roughly of about half water ice, with the rest being mixes of rock and various carbon- and nitrogen-based compounds (methane ice, ammonia ice, organic molecules, and the like). In general, their surfaces have a uniform dark gray color, but evidently they have had very different histories. All rotate synchronously about Uranus, always keeping the same face toward the planet and, just as in the case of Uranus itself, only their Southern Hemispheres were sunlit during the Voyager 2 flyby.[15]

The outermost and largest of the Satellites, Oberon, has a heavily cratered surface that is apparently quite old. Evidently little has happened to this body since the era of heavy meteoric bombardment shortly after the formation of the Solar System. There are, however, a few exceptions. Some crater floors are covered with dark material that appears to have been erupted from the interior; the two most prominent such patches occupy the interiors of craters that also have relatively bright ray systems. There are also traces of what may be long fault scarps, or cliffs, probably due to tectonic activity at some time in the moon's history. Even though Oberon's surface layers are icy, there is a great deal of vertical relief resulting form geologic activity that is probably billions of years old. The explanation for this fact is that at Uranus's distance from the Sun temperatures are so cold that water ice has a great deal of internal strength and behaves much as rocks such as granite do on Earth.[16]

The history of the next satellite inward, Titania, is very different from that of Oberon. Its most prominent features are enormous (in length, width, and depth) sinuous fault systems and canyons, evidently the result of large-scale motions of portions of the moon's crust sometime in the past. A possible explanation of these features is that the satellite's outer regions were stretched when the icy interior froze and expanded. There are many impact craters on the moon, as well as areas with fewer craters where some mechanism (perhaps "lava" flows of liquid water) resurfaced certain areas after the heaviest bombardment had ended. In general, the surface of Titania is younger than that of Oberon, and as the fault features cut across most craters, and have few crater superposed on them, they must be among the youngest features on the satellite. There are bright patches in a few places along fault scarps, evidently exposures of subsurface material that has not been darkened. The youngest features of all are bright craters with ray systems that have been produced by impacts over the past few billion years.[17]

Closer to Uranus, Umbriel is the darkest of the "big four" satellites and shows less evidence of water ice in its infrared spectrum than do the other three moons. Images revealed a densely cratered surface that, along with that of Oberon, appears to be the oldest of any of Uranus's major satellites. Moreover, essentially the entire surface is a remarkably uniform dark gray, as if it had been spray painted, though there are a few brighter spots, the most prominent being a doughnut-shaped feature that might be the floor of a large crater.[18] In general, Umbriel appears to be a body on which little has happened since its formation.

Ariel is the closest of the four largest satellite to Uranus, and its surface turns out to be the most complex of the quartet. Large parts of the surface are heavily cratered and thus probably relatively old, but there is an extensive network of younger faults, fractures, and deep valleys that evidently formed over an extended interval, as different features have been affected to various degrees by impact craters. In addition, there are smooth regions that have relatively few craters, and presumably are solidified flows of what was once possibly liquid water, water-ammonia mixtures, or other substances. In this case, too, the extensive fracturing probably occurred when a liquid interior containing water froze and expanded. The youngest features on this moon are bright spots due to impact craters and their associated ray systems that have accumulated over the past three to four billion years.[19]

In addition to any original heat left over from these satellites' formation, tidal effects may have heated Ariel at some time. Moreover, all four large satellites have densities that are greater than those of Saturn's icy moons, indicating greater percentages of rocky material, and thus greater amounts of heating from naturally radioactive isotopes. This later source may have provided the energy to produce the "lavas" that have resurfaced parts of some of Uranus's attendants.

Miranda, the smallest and innermost of Uranus's major satellites, provided the biggest surprise of them all. Though this moon is only some 500 kilometers in diameter, its surface is divided into two drastically different types of terrain. Part of the body is gently rolling and peppered with impact craters, but there are three large, relatively young areas that can only be described as bizarre (resembling to Charles Kohlhase a "prizefighter's face"). Two of these are roughly oval in outline and resemble huge racetracks that have been "plowed" on a gigantic scale, leaving behind concentric rows of high ridges and deep valleys surrounding central cores containing jumbled arrays of intersecting ridges and valleys. The third region has been dubbed the "trapezoid," after its shape, and features ridges and valleys that form sharp corners. There are also long fault scarps and valleys of various ages, with down-faulted valleys as deep as 10 to 15 kilometers, and at one location sharply edged sheer cliffs soar many kilometers more or less straight up. The younger regions display most of this moon's brightness variations, though bright regions are associated with some impact craters and fault scarps.[20]

Explaining the different histories of Uranus's major satellites clearly presents a daunting challenge. Speculations have included the suggestion that some or all of these bodies have been disrupted and then reaccreted one or more times in the past, but there are currently no certain answers. Lacking any firm ages, speculations seem destined to remain as such until more definitive answers can be obtained. Voyager had answered many old issues about the Uranian systems and yet posed many new questions.

As Voyager 2 coasted out beyond Uranus and headed for its encounter

Triton's surface, taken by Voyager 2 on August 25, 1989. (NASA photo PIA-01538)

with Neptune some three and a half years later, the spacecraft was in reasonably good shape: the suspect scan platform had performed flawlessly when operated at slow angular rates, the new attitude-control software and active "target tracking" had reduced image smear during long exposures, improved data compression techniques had worked well, the thermoelectric power supply was holding up as expected, the spacecraft as a whole had shrugged off the decreasing temperatures as its distance from the Sun increased, and—a prudent concern up to this point—there was plenty of hydrazine propellant remaining for the attitude-control system to conduct not only the Neptune flyby but also an extended interstellar mission after that. Voyager 2's success owed much to the very intensive contingency planning that had become a part of the program during the design and construction and throughout flight operations.

Rex Wade Ridenoure, a "second generation" Voyager managing engineer, who joined the Voyager team in December 1986 was among those planners. A 1976 graduate of the University of Iowa, and a champion gymnast, Ridenoure took an advanced degree in aerospace engineering at Caltech and went to work for Lockheed Missiles and Space Corporation in the spring of 1980 to work on what would eventually become the Hubble Space Telescope program. In 1985 he left to go to Utah to work with a space firm developing a basketball-sized science satellite and, a year later, responded to a call from his friend William Cosman to return to Pasadena to work on Voyager planning with Kohlhase's group.[21]

Following the Uranus encounter, Kohlhase and his staff in the Mission Planning Office focused the project's contingency planning efforts on the anticipated events of the Neptune encounter, still some three years in the future, and on the interstellar mission that would follow that encounter. For Voyager 2, the work included the development of preferred and optional flyby trajectories, the definition (and redefinition) of mission science objectives in counterpoint with very real mission constraints, and the establishment of trade-offs and options for science and engineering events. Contingency planning was a challenging and invigorating but mind-numbing process. The enormous spans of time and distance, the almost infinite number of contingencies (things that could happen), confronted by the more finite reality of electrical and mechanical apparatus and power supplies, gave contingency planning the flavor of a game of celestial chess.[22] Not the least part of the equation were the increasingly limited resources such as staff, money, and time dedicated to Voyager.

Among the "new" conditions peculiar to the Neptune encounter would be the low intensity of sunlight and the round-trip time delay and diminished strength of the radio signals. Neptune is 50 percent farther from the Sun than Uranus, so the intensity of sunlight at the more distant blue planet would be roughly half of that at the green planet, requiring longer exposures. Concurrently, the spacecraft's radio signals as received on Earth would be only half as strong as at the previous encounter and would take about eight hours for a round trip.

As in the case of the Uranus encounter, actions could be and were taken

211

here on Earth to compensate for the diminishing strength of the radio signals. Receiving antennas were enlarged, more antennas were added to the Deep Space Network, and antennas were connected in more extensive arrays that provided the equivalent of a much larger single "dish." There were also a number of small improvements in areas such as alignment, calibration, and the like, but every little bit helped.

As far as Voyager 2 was concerned, it was obviously impossible to "bring it into the shop" for an overhaul, but the spacecraft's computers could be reprogrammed from Earth. That capability had played a vital role in the Uranus encounter, and would be needed at Neptune as well. "Data compression" techniques were used in both cases and worked well. However, it is important to remember that there is no free lunch in this or any other situation; some information is thrown away, and just what you throw away depends on certain assumptions that you make. For example, instead of transmitting the actual light intensity level of each picture element (such as a "dot" on your television screen), one can transmit just the difference between two adjacent "pixels," reducing the number of data "bits" that you have to transmit by half or more. Practically, there will usually be little difference in an image where the contrast is low, but there are problems with images containing large differences in brightness.

On another front, neither the original spacecraft nor ground-based software programs could handle the long exposures expected at Neptune, for mission planners had neither the time, money, incentive, nor optimism to consider that question at earlier times. Now, however, changes were made on Earth and aboard Voyager 2 that permitted much longer exposures.

To reduce image smear during long exposures, target tracking by slowly rotating the entire spacecraft had worked well during the Saturn and, especially, Uranus encounters. For Neptune, there was a new wrinkle. The previous technique cut off direct communication with Earth for a short time, because the spacecraft's antenna was pointed away from our planet while a series of images was taken. The new method, dubbed "nodding," let the probe's antenna remain pointed fairly close to the direction of Earth while an image was taken, then repositioned it back to a direct "point" at Earth. Of course the camera next had to be repointed for the next image in the series. Making all this work required very short and precise firings of the attitude-control thrusters. Finally, there was the possibility of moving the scan platform itself; however, for a variety of reasons (the scan platform's actuator had been suspect ever since the Saturn encounter) this capability would be used only for fast moving targets that could not be imaged in any other way.[23]

If little was known about Uranus before Voyager 2's visit, scientist knew even less about Neptune. Although its mass was well determined, thanks to the observed motions of its satellites Triton and Nereid, its diameter was only roughly known, as was its average density, which was somewhat greater than that of Jupiter or Saturn, much less than those of terrestrial planets such as our own, and in the same range as that of Uranus.

Neptune's axis of rotation was known to be tilted roughly about the same amount as Earth's, and the rotation of its upper atmosphere was estimated at somewhat under eighteen hours. No cloud bands or other permanent or long-lived distinctive features had ever been discerned from Earth, though in the years immediately preceding the Voyager 2 encounter there were credible observations of distinct, probably variable cloud features. Still, these were only vague hints that provided little concrete evidence of what Voyager 2 might reveal.

The rotation period of Neptune's interior was unknown. There had been no detection of impulsive, periodic radio emission and, although there were some reports from Earth observations of periodically varying radio emission at some wavelengths, it was uncertain just where and how this radiation was produced. Although it was generally assumed that Neptune had a magnetic field, there was no evidence for its existence and, of course, no substantial hints whatever as to its strength and orientation.

Neptune's satellites were mere points of light in terrestrial telescopes, though Triton was known to be one of the largest moons in the Solar System. This giant satellite was believed to possess a thin atmosphere, and there were firm indications of frozen or gaseous methane in its spectrum. In a bizarre twist, Triton is the only large satellite of a major planet that orbits in a retrograde direction, that is, opposite to the direction of the planet's rotation.

Nereid, in contrast, orbits in a direct mode, though its large orbit is extremely eccentric; in telescopes it is a mere point of light and nothing

Irving Reed and Gustav Solomon, developers of the Reed-Solomon encoder. (Courtesy of
Charles Kohlhase)

more. In short, astronomers knew little about Neptune and its system as
Voyager 2 approached. However, as in the case of Uranus, there was one
spectacular exception: Neptune had rings. Or did it?

After rings were found around Uranus and Jupiter, Neptune was the
only giant planet that was not known to have such a system, so naturally
observers set out to see if this was really the case. Clearly any rings could
not be as prominent as those of Saturn, or they would have been seen long
ago, but they might be very tenuous, like that of Jupiter, or thin and dark,
such as those around Uranus. The method of stellar occultations had done
the job in the case of Uranus, but when it was tried on Neptune the results
were puzzling indeed.

By the time of the Voyager 2 flyby, about fifty occultations had been observed from Earth by various observers, yielding just over one hundred
separate "scans" across the Neptune system. Most of these data sets
showed only the extinction caused by the planet itself, but a in a few cases

(at least five, perhaps as many as eight) showed something else: momentary sharp drops in the brightness of the background star either before or after the planetary occultation. A complete ring would generally cause dips both before and after the main occultation, so what was going on? Moreover, why did most of the observations show no trace of material in orbit around Neptune?

It was possible that these puzzling observations were caused by one or more small satellites, or that they were due to transient rings. However, the most popular explanation involved a narrow, incomplete ring or rings; that is, discontinuous rings or arcs orbit Neptune. Hopes were high that Voyager 2 would clear up this situation.[24]

As Neptune was the last known planet that Voyager 2 would encounter, mission planners had the luxury of choosing the flyby trajectory that promised the greatest scientific returns. The principal aims were (1) a close approach to the giant satellite Triton, along with occultations of both the Sun and Earth as viewed from the spacecraft; (2) a close pass by Neptune, also including Sun and Earth occultations; and (3) ensuring that the occultations took place when Neptune was relatively high in the sky as seen from the Australian tracking station. The last requirement was important because at the time of the encounter, Neptune was south of the celestial equator in Earth's sky and so, as was the case with the Uranus encounter, best observed from the Southern Hemisphere. Mission planners also struggled with the problems of how close Voyager could approach Neptune and still miss the ring arcs, atmosphere, and anticipated bands of radiation.[25]

The selected trajectory targeted closest approach to Neptune for August 25, 1989, at a distance of some 29,000 kilometers from the planet's center, with closest approach to Triton about five hours later (there would be no close approach to Nereid). During the flyby, radio signals from the probe would take four hours and six minutes to reach Earth.[26]

As early as January 1989, Voyager 2 images of Neptune revealed a bright discrete cloud feature that stayed around, even though its shape changed constantly. In the next few months, features came and went, but several more long-lasting features came to light: bright, dark, and bright and dark. Thus, months before closest approach, it was clear that the top of Neptune's atmosphere would present a much more complex and interesting appearance than that of Uranus.

Neptune turned out to possess a clear atmosphere on top of a uniform, almost featureless, planet-shrouding cloud layer possibly composed of hydrogen sulfide or ammonia ice crystals. There were a few vague, alternating belts of relatively brighter and darker zones at different latitudes, but nothing as distinctive as those on Saturn, let alone those on Jupiter.

As the resolution of Voyager 2 images improved, it became clear that the most prominent white feature was a complex of high clouds around the southern edge, and sometimes partially over, a large, dark, oval feature dubbed the Great Dark Spot by analogy with Jupiter's Great Red Spot. This dusky feature was roughly the size of Earth, lay at a latitude of twenty degrees south, and rotated with a period of about sixteen days in the direction of an anticyclone (counterclockwise in the Southern Hemisphere). Though the spot was evidently lower than the general cloud deck, it acted as a barrier to the prevailing winds, forcing them upward and causing condensation of "lee" or "orographic" clouds such as mountains sometimes produce on Earth (and Mars, for that matter), though in this case the condensed material is presumably crystals of methane ice. High-resolution images supported this explanation; the white feature broke up into separate narrow cloud streaks, and, though the general appearance of the complex changed only slowly, small individual features passed rapidly through it, just as in terrestrial cases.[27] With the benefit of hindsight, it was possible to trace the bright, "edging" cloud feature back through some five years of Earth-based telescopic observations. Presumably, that means that the Great Dark Spot had been around at least that long.

Another persistent bright feature lay near latitude forty-two degrees south and was dubbed the "Scooter." At high resolution it appeared as a cluster of narrow, parallel light streaks and seemed to lie at a lower level than the other bright, cirruslike clouds. The overall shape of this feature varied greatly, but it was present during the entire interval that Voyager 2 images were able to resolve it.

Another type of long-lasting feature, again in the Southern Hemisphere, was designated "D2"; it first appeared as a dark spot and then developed a bright core containing complex cloud features. Yet another different kind of bright feature was actually a long, drawn-out region of variable cloudiness in the polar region at seventy-one degrees south. This area was narrow in the north-to-south direction but extended a quarter of the way around the planet in the east-west direction; cloud formations here changed significantly from Neptunian day to day.

There were many other bright, transient cloud features, most of which were probably composed of methane ice crystals floating some 50 to 100 kilometers above the solid, underlying cloud deck. In some cases, the altitudes of these high-level clouds could be directly determined from measurements of the shadows that they cast on the cloud blanket below.

Voyager 2 observations of variable radio emission from Neptune and its vicinity revealed a periodicity of 16.11 hours (roughly 16 hours, 7 minutes), but it proved to be a challenge to determine the rotation periods at different latitudes in Neptune's atmosphere. A few persistent features such as the Great Dark Spot provided precise rotation periods at a few latitudes, but most of the bright clouds came, went, and changed so rapidly that it was difficult to decide if a feature seen during one rotation of the planet was the same as that seen on the next. Despite this obstacle, it was possible to determine that the upper atmosphere of Neptune rotates more slowly near the equator than it does near the South Pole, which is the opposite of the cases of Jupiter and Saturn.

Neptune is a windy place, having atmospheric currents with speeds with respect to the planet's interior approaching that of sound; in fact, it may well be the windiest place in the Solar System. Some or all of this rather violent activity may be due to the fact that Neptune, unlike Uranus, has a substantial source of internal heat, giving off almost three times as much energy as it receives from the Sun. Given this difference, it is surprising that the pattern of atmospheric motions on the two planets is so similar.

Voyager 2 also revealed that Neptune does indeed have a magnetic field, as expected, but what was not expected was that the equivalent "magnetic dipole" was not only tilted some forty-seven degrees to the rotation axis but also displaced more than half the distance from the planet's center to its "surface." The fact that two planets have similarly bizarre configurations of their magnetic fields means that it is unlikely that the orientation of Uranus's field is due in some way to its unusual axis of rotation. Moreover, the chance that we have just happened to catch both of these bodies in the act of reversing their magnetic fields (as happens periodically on Earth) is most improbable.

In related matters, there were indications of weak aurorae on Neptune, but no evidence of lightning either in direct images or from radio observations.

As was hoped, the ring situation was cleared up, Voyager 2 finding four (or more) rings, two exceedingly narrow and well defined, and two tenuous and relatively wide in radial extent. The most prominent annulus was the outermost, a wire-thin structure that displayed three denser segments, all within a thirty-three-degree segment of its circumference. These segments evidently were responsible for almost all of the occultation events observed from Earth. Inside that ring was an extended, diffuse feature, and inside of that another narrow, unresolved ring. Inside of all of these was another diffuse, extended ring. In addition, there were indications of an extremely tenuous dust sheet extending throughout the ring system, and hints of structure in that sheet as well as other features of uncertain nature. Representative ring particles are somewhat smaller than those in Uranus's system, say grains of sand compared to soccer balls, and all are extremely dark.[28]

The rings were not readily detected by the radio occultation experiment, and the optical occultation of the star Sigma Sagitarii gave a width of 10 kilometers for the core of the outermost ring. Some of Neptune's rings seem to have relatively more dust than the main ones of Saturn and Uranus, as in the case with the arc segments of the outermost ring. The very existence of the arcs is a puzzle, for straightforward application of the laws of celestial mechanics suggest that they should be smeared out around the ring's orbit in only a few Earth years.

One possibility for confining the ring arcs, as well as producing narrow rings and sharp ring edges, is that of small shepherding satellites; as in other cases, many were looked for but few were found. Voyager 2 discovered six new, small, relatively nearby satellites of Neptune; all are irregular in shape, have dark surfaces, and move in circular, direct orbits in the planet's equatorial plane. With surprisingly large diameters that typically measure hundreds of kilometers, these bodies are, as a class, much larger than, say, the small inner satellites of Uranus. Four of these bodies move within Neptune's ring system and two outside, but these were not nearly enough to do the job. Moreover, a diligent search of the ring system and its vicinity turned up no additional satellites down to a detection limit of about 12 kilometers in diameter. Of course, as in the cases of Saturn and Uranus, it is possible that even smaller moonlets could do the job and yet remain undetected. Still, the complete lack of evidence for any other satellites was discouraging to those who championed the shepherding satellite theory.

Triton was a complete surprise to every space scientist. Although this satellite currently moves in a circular orbit above Neptune's equator, its retrograde motion (with respect to the planet's rotation) strongly suggests that it formed independently and was captured at some time in the distant past. Astronomers generally assumed that tidal forces had long ago forced Titan to rotate so that it always presented the same face to Neptune, a surmise that turned out to be correct. On the other hand its diameter, 2,700 kilometers, turned out to be smaller than that estimated from Earth, because its surface is, on average, about the most reflective of any body in the Solar System. Because this moon reflects almost all the incident solar energy back into space, its surface is the coldest of any yet measured among the Sun's family. There is a thin atmosphere, the surface pressure of which is roughly seventy thousand times more rarefied than Earth's and is composed of molecular nitrogen with a "whiff"—about one part in ten thousand—of methane.

The atmospheric pressure and temperature at Triton's surface are close to the triple point of molecular nitrogen, where that substance can exist in solid, liquid, and gaseous states, similar to the role that ordinary H_2O plays on Earth. Voyager 2 determined that this satellite's density was about twice that of water, and presumably it is composed of a combination of rock and ices, with ordinary water ice probably dominating.

There is substantial vertical relief on Triton, even though most of the elevation differences occur in water ice, which behaves as a strong rock at very low temperatures. The high and low areas, and indeed essentially the entire surface that Voyager 2 imaged, appears to be covered by a relatively thin, fresh layer of ice or snow, probably composed of solid nitrogen, along with some methane.

Triton's surface presents a bewildering array of what can only be called bizarre landforms. There are long, sinuous ridge and valley systems; smooth, flat plains due to multiple flooding by some sort of "lava" (probably water); areas that resemble the skin of a cantaloupe; and regions that defy simple description. One type of terrain that was not seen was any old, heavily cratered surface. Although different regions on Triton have different densities of impact craters (though there are some regions whose appearance is so complex and unusual that it is uncertain if they host any craters at all), it appears that the regions surveyed by Voyager 2 are young in terms of the age of the Solar System.

Portions of Triton's surface look so young that, contrary to all thinking before the encounter, there is a distinct possibility that the satellite is currently active geologically. This improbability became a certainty when images revealed two geyserlike plumes that spewed dark material some eight kilometers up into the satellite's tenuous atmosphere, where the plumes flattened out and were blown downwind some 100 kilometers by upper-level winds. Fallout from eruptions such as these provides a reasonable explanation for the many dark streaks seen on the moon's surface, but there are no certain explanations of just what the dark material is, or just how these geysers are powered. This puzzle is just a part of a larger question: How can a body smaller than Earth's Moon, with no obvious energy source such as tidal interactions, remain geologically active to the present day? There are almost as many speculative suggestions as scientists interested in the subject, but no certain or even generally accepted answers.

TERMINATION SHOCK

Following the close flybys of the Jupiter and Saturn planetary systems by Voyagers 1 and 2, Voyager 2 completed flybys of the two outermost gas giants, Uranus and Neptune, and in 1989, twelve years after launch and almost 3 billion miles (8.3 billion kilometers) from the Sun, the twin spacecraft began a new mission. The Voyager's Grand Tour of the outer planets had been accomplished, and there began yet another new epoch of Voyager explorations, the Voyager Interstellar Mission. This marked the beginning of NASA and humankind's exploration beyond the Solar System and the beginning of the journey into interstellar space.[1]

Some astronomers believed that the most spectacular discoveries were yet to come, for like the Voyagers themselves, "the Sun and its family of planets are wanderers in our Galaxy and the journey is through a sea of primordial matter left over from the birth of the Universe."[2] There was and still is much to learn about interstellar space and the boundaries defining the Solar System.

Astronomers have argued for centuries about where the Solar System ends, and there is still no consensus on an answer. One opinion is that the effective boundary is roughly halfway to the nearest star, beyond which the Sun's gravitational influence is less than that of another luminary. This distance is estimated to be about 100,000 times greater than that from Earth to the Sun. Should that be the case, there is no hope of either Voyager getting that far in working order.

But there is a more definitive characteristic of the Solar System's boundary that is unambiguous, and which the Voyagers will most likely approach or pass through—namely, the dividing surface between the solar and interstellar winds. The solar wind is a tenuous flow of charged particles, of predominantly negative electrons and positive protons, the two components of the hydrogen that makes up most of our own star. This phenomenon was discovered by spacecraft at the very beginning of the space age in the late 1950s and early 1960s and consists of particles literally evaporating from the Sun's intensely hot corona, or outer atmosphere, as well as those spewed out by energetic events such as solar flares (which themselves are poorly understood but are violent enough to initiate nuclear reactions in some cases). At that time this wind was estimated to extend out to somewhere not far beyond Jupiter's orbit.

The solar wind has to stop somewhere, for astronomers have detected interstellar winds of charged particles. These are produced by such mechanisms as gigantic supernova explosions, where stars literally blow up with catastrophic results; by the pressure of light from extremely hot, luminous, and massive stars; and by evaporation from the outer atmospheres of very hot stars of assorted types.

Theoretical estimates of the distances from the Sun to which the solar wind extends grew and grew as the Pioneer and Voyager spacecraft cruised outward and found no evidence of any end to the outflow. This situation often develops when theory bumps up against hard data. Still, with the solar wind blowing outward, and the interstellar winds blowing against it, there must be a boundary out there somewhere. The attempt to find that dividing surface became an important objective of the Voyager Interstellar Mission.

Paradoxically, the first indication that the solar wind is reaching its limit will be a dramatic slowdown in the speed of the particles that compose it, along with an equally dramatic change from a more or less outward flow to a chaotic, turbulent state—the termination shock. Odd as this change in behavior sounds, Ed Stone suggests a simple yet technically correct example of this phenomenon that any reader can easily perform. Just take a large dinner plate or board and place it in the bottom of a sink, then position the spout over the center of the flat surface and turn the water on fully. When the stream hits the surface, it will move smoothly outward, radially and in a relatively thin, uniform layer. At some point, however, the motion will change to boiling, roiling, turbulent flow that moves in many directions at once, and incidentally is much thicker in the vertical direction. The partition between the two types of motion is a true terminal shock.[3]

Beyond that shock surface, at some unknown distance, but probably taking the Voyagers several years to traverse, lies the edge of the heliosphere, beyond which is the domain of the interstellar winds (In the kitchen sink example, the edge of the plate or board determines that.) If either of the Voyagers get so far, they will be able to provide our first direct information about the character of "normal" matter from beyond the Solar System. Until now, all scientists have had is electromagnetic radiation (gamma-ray, x-ray, ultraviolet, visible, infrared, and radio waves) and cosmic rays, these last being charged particles, but ones that evidently have had a rather unusual history. One of the questions that the probes might answer is the relative proportion of helium to hydrogen nuclei; another is the abundance of heavier nuclei, such as carbon, nitrogen, and oxygen. Moreover, the possibility of discovering more massive particles, such as primitive organic molecules, is an unlikely yet fascinating possibility.

NASA scientists assigned the new Voyager Interstellar Mission four basic scientific investigations, including (1) the search for the origins of life, (2) a study of the structure and evolution of the universe, (3) exploration of the Solar System and outer planets, and (4) a better understanding the Sun-Earth connection. But performing these investigations faced rising new obstacles.

The Voyager flights beyond the outermost planets of the Solar System en route to interstellar space could occupy as much or more time and distance as did the Grand Tour itself. And as time and distances lengthened, power reserves, fuel supplies, and communications became increasingly critical factors, and back on Earth financial support waned and staffing became more and more difficult. The passage of time, coupled with the seemingly diminishing scientific returns, declining media and public interest, and new NASA programs competing for funding—not to mention the basic engineering and technical problems related to space exploration at extreme distances and in a hostile environment—combined to make the future of the Voyager program increasingly tenuous.

But there were still powerful incentives to continue the work. Already, as early as May 1993, there were intriguing hints that the interstellar winds are not too distant from the Voyagers. The spacecraft began to detect low-frequency radio emissions that appeared to come from somewhere beyond the Solar System. The best explanation for this radiation is that it is produced when clouds of particles in the solar wind collide with one another along the surfaces created by interaction with the interstellar winds. These radio emissions are unexpectedly energetic, but their low frequency means that they cannot penetrate our ionosphere and so cannot be detected from Earth. In addition, energetic charged particles presumably produced at the collision sites between the Solar and interstellar winds have been detected.[4]

Yet even as Voyager 1 moved toward the edge of the Solar System there was yet one final opportunity to accomplish an astronomical observation never before possible: to photograph our planetary system from the outside looking in. Previously such an undertaking had been limited merely to the imaginations of astronomers and science fiction writers, but now the probe was just far enough from the Sun, and in the right position with respect to the major planets, to actually produce such an image.

This was to be the final assignment of the probe's imaging system, but to produce a photo montage of the Solar System would be a difficult task. The spacecraft was so far from the Sun that the terrestrial planets appeared very close to our star, posing a very real threat of "burning out" the television cameras if they were not pointed accurately. JPL's scientists and engineers succeeded in delivering the radio signals via an eight-hour journey through space instructing the scan platform to point in such a way that the cameras did indeed capture the Solar System from the outside looking in, producing a remarkable planetary array.[5] It was another first for Voyager and for humankind.

Once this final backward look was taken, controllers on Earth turned the cameras off on Voyager 1 as had been done earlier with Voyager 2, to conserve electrical power. The moveable scan platforms were left functioning for some time to permit pointed ultraviolet observations of specific celestial bodies far beyond the Solar System, such as hot stars and active galactic nuclei, then these systems too were shut down. Eventually, newly launched Earth satellites gradually eliminated the need for that function, and later both Voyager's scan platforms were deactivated in order to save electrical power and hydrazine thruster fuel. Nevertheless, random, "unpointed" ultraviolet observations of the sky continued for some time from both probes.

Over the next few years various systems and subsystems on both probes were shut down one by one to extend the lifetimes of those remaining. Not doing any "pointed" observations, whether with the scan platform or by reorienting an entire spacecraft, saved hydrazine, and electrical power was conserved by turning off heaters that kept the temperatures of various mechanical and electrical components within their operating limits. One casualty of switching off the heaters was ultraviolet observation.

Yet even in their powered-down condition, both Voyagers carry substantial arrays of operating instruments capable of returning scientific data about charged particles, magnetic fields, and radio emissions, as long as the spacecraft remain alive. Each vehicle contains a cosmic ray experiment (for high-energy charged particles), a low-energy charged-particle experiment (for studying solar and interstellar winds), a magnetometer to measure the weak magnetic fields between the stars, a plasma subsystem, a plasma-wave instrument, and a radio astronomy experiment. Should they make it that far, these instruments would provide scientists with previously unattainable information about interstellar space.[6]

On February 17, 1998, Voyager 1 passed Pioneer 10 to become the most distant artifact ever launched from Earth. The Pioneer 10 and 11 exploratory missions and their vital functions had ended while the spacecraft pressed on, active and alive but still far from interstellar space. The first indication that the Voyagers were nearing interstellar space would involve the anticipated encounter with the termination shock, which is projected to occur in the first decade of the twenty-first century. At that time scientists expect Voyager 1 to be about 7 billion miles (over 11 billion kilometers or 85 AU) from the Sun and traveling at about 39,000 miles per hour. In January 2000 the Voyagers were at approximately 80 AU, and by 2003 the range is estimated to be 90 AU.[7]

But termination, according to Ed B. Massey, who became JPL's Voyager project manager in 1998, could be due to things of Earth rather than to the physical laws of science, space, and technology. Massey, a native of Alabama and graduate of Tuskegee Institute, received his commission in the U.S. Air Force and spent much of his career in missile defense and satellite tracking work before retiring in 1986. He joined JPL's Ulysses Project team in 1987 and became project manager in 1996. Following Voyager's last planetary encounter, Massey noted, budget and staff reductions became the rule.[8] The termination of the Voyager program might well relate to administrative and financial concerns and may well precede the actual demise of Voyager operating systems, expected to occur around 2020.

In 1998, as an economy measure, Voyager operations and administration was folded into the Ulysses Project, and Massey became the manager of both programs. By the close of the century Voyager mission funding had dwindled to $5 million annually, and the science and operations staff stood at sixteen or seventeen, with further downsizing expected. Nevertheless, in technical terms the missions were still very much on course. Science instruments were recording constant changes in cosmic rays and energetic particles, as well as the velocity of the solar winds, and magnetic field readings were falling to the lowest levels ever seen. Meanwhile, the RTG power supplies were holding out better than expected. Power, Massey explained, though critical to the continuance of the Voyager missions, would not be the determining factor in the termination of the project.[9]

Rather, science returns on Earth were being measured against costs and public and political support. The Science Steering Group, in its annual meetings, strongly supported continuance of the Voyager missions. The Sun-Earth Connection Review Board, which provides a degree of umbrella oversight for NASA's diverse space exploration programs, gave Voyager overwhelming approval, while the decisions of a NASA Senior Review Board would have much to do with future operations.[10] Science returns were also being measured against other projected space science investigations.

The science community and NASA were already initiating and formulating new Solar System and interstellar exploration programs justified in good part by Voyager discoveries. For example, Galileo, launched in 1989, began a two-year orbital survey of Jupiter in December 1995, followed by an extended study of the planet's environment. By December 1999 the vehicle had completed a four-and-a-half-year follow-up mission intensively investigating Jupiter and its four largest moons. These explorations have indicated that Io likely has over three hundred active volcanoes, far more than the eighty-one located by Voyager, and that new eruptions are frequent. One volcano, Loki, is apparently the most powerful in the Solar System. Io is a distinctly nonterrestrial volcanic experience.[11]

Ulysses, launched in October 1990, began an intensive orbital survey of Jupiter and its environs in 1992, and Mars Surveyor and lander programs absorbed considerable NASA funds and personnel throughout the eighties and nineties. Cassini, launched in 1997, began a new and more intensive exploratory mission to Saturn and its largest moon, Titan. This probe followed a complex trajectory, including two close flybys of Venus and one of Earth, in order to pick up the gravitational assists needed to reach its final destination. A close Cassini flyby of Jupiter late in 2000 provided a realistic dress rehearsal for operations at Saturn. Finally, the spacecraft is scheduled to orbit Saturn in 2004 and drop an entry probe into Titan's atmosphere in 2005.[12]

Deep Space 1, launched in October 1998, is testing new equipment and technologies to help provide quality missions at lower costs, and Stardust, launched in 1999, is designed to collect dust samples for Comet Wild 2 and interstellar space. All of these missions, though stimulated by earlier Voyager investigations, came to compete with Voyager for time, attention, and funding, even while a broader context other space flight programs, the space station, and orbital observatories vied for NASA and national resources. Still, many of these efforts are a testament to the remarkable successes achieved by Voyager and its continuing impact on planetary exploration.

Thus, by any measurement, at the beginning of the twenty-first century Voyager had become one of the most distinctive and successful exploratory missions of the twentieth century. It was a triumph in innovative engineering, a "STAR" (self-test and repair system) of sorts, that, with help from a diligent ground crew, was constantly reengineered as it flew through time and space, producing scientific data that changed our basic knowledge of the Solar System.

The Voyagers have demonstrated in dramatic style that there are more than just the two kinds of planets that astronomers earlier had categorized—that is, the relatively small rocky bodies with relatively thin atmospheres such as Earth, and the monstrous "gas giants" such as Jupiter. Instead, the Voyagers revealed the existence of an entirely new class of bodies, the "ice worlds," not to mention such bizarre objects as vigorously volcanic Io and Europa with its worldwide ocean covered by a worldwide sheet of ice and the possibility of some sort of primitive life in that dark sea.

For over a quarter of a century the Voyager spacecraft have provided information heretofore inaccessible and largely unanticipated. Even by the close of the Neptune encounter some five trillion bits of scientific data had been returned to Earth by the twin explorers. Signals collected by the deep-space tracking antennas even then were so weak that a digital watch of the time generated a power level twenty billion times greater than that of the incoming radio signal. Then, too, the navigational accuracy of the flight to Neptune is equal to the feat of sinking a 2,260-mile golf putt.[13]

Voyager is a distinctive engineering and scientific achievement, but it is also more. It is an accomplishment of team engineering that drew from the essence and spirit of NASA, the Jet Propulsion Laboratory, and the American people. It is itself technologically, financially, and by inspiration and definition the creation of history and of the accumulative space technology of the twentieth century. Voyager is a technological composite of the experiences and knowledge of the Jet Propulsion Laboratory and of NASA's Ranger, Corporal, Mariner, Viking, STAR, TOPS, and Grand Tour programs.

Moreover, the Voyager program is an intergenerational effort. Few, if any, of those who designed, built, and launched the probes will live to see their final days, estimated to be at the close of the second decade of the twenty-first century. Many of those who managed and operated Voyager flights after launch had little or no knowledge of the program's inception and development, and many scientists of the twenty-first century who will use data derived from the missions were not born when the Voyagers were launched in 1977.

Interestingly, and suggestive of the future character of space flight and exploration, during the quarter-century and more since launch, affairs on Earth underwent a sea of change. Jimmy Carter was inaugurated as president of the United States in 1977, the year Voyagers were launched. In that same year, the Organization of Petroleum Exporting Countries

(OPEC) embargo drove the price of oil up from two to thirty-two dollars per barrel and Alaska North Slope Oil was just coming to market. Concorde jet service to London and Paris had begun, and the United States concluded a treaty with the Republic of Panama granting the latter sovereignty over the Panama Canal. That same year Steven Jobs and Stephen Wozniak incorporated the Apple computer company.

At the time of the Voyager 1 encounter with Jupiter, In November 1980, Ronald Wilson Reagan was elected president. In the following years the United States resumed a military buildup, experienced a recession, and began the Strategic Defense Initiative using so-called Star Wars technology. Reagan was reelected president in 1984, and Mikhail Sergeyevich Gorbachev became secretary general of the Soviet Union's Communist Party in 1985, implementing policies of glasnost and perestroika, peace and reconciliation, the beginning of the end of the cold war.

George Bush won election as president of the United States in 1988, and the next year, at the time of Voyager 2's Neptune encounter, the Berlin Wall dividing the Communist and Western worlds opened and was soon dismantled—as was the Union of Soviet Socialists Republics. As the Voyagers approached the heliopause, McDonald's opened a restaurant in China's previously closed capital city of Beijing (1992), denoting a markedly changing relationship between the United States and that country. William Jefferson Clinton became president-elect of the United States in 1992, and Congress approved the North American Free Trade Agreement (NAFTA) in 1993. The following year Congress approved the Uruguay Round of the GATT (General Agreement on Tariffs and Trade) joining 124 nations in an almost revolutionary trade agreement. Even as the Voyagers sped outward creating a greater awareness on Earth of the integrity and interrelationship of the Solar System, the planet which it left behind a quarter-century earlier was being transformed into a new, more integrated community. And there was, some believed, a causal relationship.

Thus, Steve Maran, public information officer for the American Astronomical Society, considers the Voyager missions to be "one of the great milestones of human history." In like vein, David Morrison characterizes Voyager as "the greatest piece of exploration of our century." Morrison, who remembered being committed to becoming an astronomer as early as the sixth or seventh grade, graduated from the University of Illinois and completed his master's degree and doctorate in astronomy at Harvard University. His dissertation under Carl Sagan examined the temperatures of the terrestrial planets. He spent the next nineteen years at the Institute for Astronomy at the University of Hawaii, serving variously as vice chancellor for graduate education and research, and director of the NASA Infrared Telescope Facility, and briefly in 1981 as deputy associate administrator for space science at NASA Headquarters. In 1988 he went to work for NASA Ames as chief of the Space Science Division. Morrison considered Voyager on a par with the Viking missions as "NASA's greatest triumph from an intellectual point of view. Voyager transformed our view of the Solar System."[14]

As we have seen, the Voyagers returned to Earth the first pictures of the outer Solar System, and many celestial objects, previously known only as "points of light visible through a telescope," were discovered to have had complex geological histories. These bodies could now be viewed as "real worlds with a history, a geology, and a personality that was completely unknown, not even hinted at previously."[15]

Morrison agreed with most scientists that the greatest surprise was the active volcanism on Io. "We had confidently expected Io to be a heavily cratered world, lunar-like in its geology, but with a unique chemistry," he recalled. But, in addition, Voyager chemistry had as much to do with the human spirit as with celestial bodies. These missions, Morrison believes, particularly during the 1979–81 Jupiter/Saturn encounters, produced a profound level of public engagement, for there was a shared experience of discovery. Moreover, the encounters occurred during the time when Carl Sagan was airing his *Cosmos* public television series, which further galvanized public excitement.[16]

For much of their historic flights, the Voyager missions became "media events." Almost everything connected with the project was newsworthy, and press conferences during the planetary encounters were always crowded affairs. Although the public demanded "instant science," newsworthy science was almost by definition incompatible with the spirit of scientific inquiry, and, indeed, inconsistent with the character of incoming data from the robot explorers.

To give just one example, Voyager 1's encounter with Saturn generated a media response that was unprecedented in the history of unmanned space exploration. The event was the cover story for both *Time* and *Newsweek,* the first time both publications had featured a space-related subject since the initial Apollo 11 manned Moon landing in 1969. During the encounter more than five hundred news representatives converged on JPL, coming from around the world; the immediate television audience was estimated at 100 million viewers. Publicity was not confined to front-page newspaper stories and television sound bites, for many editorial and opinion pages contained laudatory articles, and later there were more extensive descriptions of Voyager 1's triumphs. Planetary exploration was definitely news.[17]

The Voyagers even made it to Hollywood. In the first *Star Trek* movie (unimaginatively entitled *Star Trek: The Motion Picture*), Captain Kirk, Mr. Spock, and the old, original crew track down an immensely powerful, enigmatic celestial entity known only as "Veeger." It turns out that at the heart of this mysterious Veeger is nothing less than an old Voyager spacecraft, highly modified by an advanced alien civilization that it had encountered during its centuries of interstellar wanderings. The probe was easily identified—for such things as an American flag, "NASA," and "JPL" were painted on its side. Once Mr. Spock had made the needed modifications to Veeger's electronic circuits, the huge and now "living" machine that could have threatened Earth was deflected and the USS *Enterprise* had once again saved galactic civilization. All in all it was a rather slow-moving film, but Voyager was an undisputed movie star.[18]

Voyager made it into the auto world as well. If it had not already been so, that name became synonymous with adventure, and General Motor's Corporation's Plymouth Voyager minivan achieved public recognition and acclaim in part because of its name association with NASA's missions. Although General Motors named its Saturn automobile after the Saturn rockets used to boost the Apollo spacecraft into orbit, the Saturn logo is a stylized version of Saturn and its ring system as could only have been viewed from the Voyager perspective. But there was, Voyager notwithstanding, a NASA/space exploration explanation to the Saturn auto and Voyager minivan names. As explained by Kelly Carey at General Motor's Saturn Customer Assistance Center,

Back then, our foreign competition (Russia) had been first to launch a satellite into space (Sputnik, in 1957) and first to launch a man into space (Gagarin, in 1961), so they definitely had the advantage in the industry of space flight.

Likewise, General Motors wanted to launch a small car project that could create vehicles, designed and manufactured in the United States, that could beat foreign competition in a marketplace which, at the time, was dominated by foreign imports. Therefore the project was dubbed Saturn from the early days, in memory of the Saturn rockets.[19]

Happily for the marriage of Voyager science and the media, Voyager science experiments had a strong focus on "pictures," which in earlier times had been generally regarded by scientists as frivolous to serious inquiry. Imaging—previously associated with intriguing popular photographs having little scientific importance, such as those churned out by Lowell and other observatories—emerged during the Voyager era as a highly productive field of study. The advent of digital video cameras using CCDs changed the situation, as photographs became not just pretty pictures, but arrays of photometric points. Thus, Voyager imaging produced quantitative data and an enormous amount of information. Bradford A. Smith, who headed the Voyager Imaging Team, ranked this technology and particularly the discoveries of the volcanoes of Io and the complex structure of Saturn's rings as the greatest Voyager discoveries.[20]

The Voyagers, he believes, are the most successful unmanned missions ever flown by NASA. Moreover, as Morrison also observed, the missions stirred huge public interest and excitement, and helped to create a better understanding of Earth as a member of a community of planets. Then, too, Voyager epitomized the new reality of science in general, and planetary sciences in particular, as a team effort. It was a far cry from the "old time stuff, the great astronomer using a telescope—one guy doing one thing." And that team science, Smith believes, was nourished at the Jet Propulsion Laboratory and fostered by the NASA space programs.[21]

David Crawford, an observational astronomer at Kitt Peak National Observatory, whose doctoral dissertation on photometric techniques with applications to galactic structure and open star clusters was completed at the University of Chicago, never worked for NASA or had NASA funding his research, but he described Voyager in a different context as a "real adventure for the soul, a real adventure for the mind and the spirit." He thought that explained, in part, the public's fascination with Voyager missions in particular and space science in general. Voyager transformed planetary astronomy from a dormant, if not dead, subject and put it "right in the core of exciting things. . . . It opened vast new windows to the universe." In this particular case, a window to the planetary system that is an integral part of the universe.[22]

Crawford suggests that one of the most important windows opened by Voyager was that of the human mind: "Most people don't look very far beyond their nose. Or beyond their own town. And we live in the universe." So many of the problems we have derive from people having no vision of being part of Earth, much less part of the universe. "The more we can get people looking out . . . the better off the Earth is going to be," he said.[23]

There was, David Chandler concurs, a growing awareness of the universe during the Voyager encounters. Beginning as a free-lance writer and then as science writer for the *Boston Globe,* he followed the missions with detached interest during the first Jupiter and Saturn flybys, but became an active participant when he was assigned to cover the Uranus and Neptune encounters. The photos from those encounters, he said, were dazzling. But the real impact of Voyager had more to do with an awakening of the hearts and minds of humankind. The emergence of what formerly had been simple pinpoints of light, not only as "real worlds" but "such an extraordinary variety of worlds, so different from one another . . . so different than anything anybody had anticipated," revolutionized thoughts and attitudes about the Solar System and Earth's place in the universe. Io and its volcanoes were a high point, as was Miranda with its incredibly varied surface, and Triton was just amazing. Voyager, Chandler concluded, "was the Lewis and Clark expedition" of the Solar System. It took a blank canvas and made it into something that began to have real details and real information in it.[24]

Voyager, at least for a time, diverted humanity from Earth-centered thoughts. And to some extent, this new awareness was a humbling experience, Ed Stone surmised. We came to realize, he believes, that nature is much more diverse than our ideas about it and can take universal physical laws and produce things that astound us. For example, planets are of diverse composition: there are ice worlds, gas worlds, and terrestrial worlds. Some of the ice worlds are charcoal black, not snow white, and, in fact, the former are widespread in the Solar System. The question is, what is this dark material? There are volcanoes on Io, and recent or past tectonic plate movements on frozen celestial bodies where the surface temperatures are 120 Kelvin or less. There may be a liquid water beneath Europa's icy crust, and there are violent geysers on Triton, the coldest body yet visited in the Solar System. In short, Voyager has been a learning experience for the scientist, the engineer, and the general public.[25]

It was, agrees Ellis Miner, NASA's most successful interplanetary mission ever flown. The most important discoveries by the Voyagers, Miner believed, really had to do with the missions' cumulative results. They confirmed that the environment of the Solar System was remarkably diverse. Instead of finding only "dead bodies," the Solar System was "alive." Io was most prominent, but some activity was discovered on Europa, Enceladus, Saturn, Titan, Triton, Neptune, and Miranda. Even Titania and Ariel, Miner noted, were far more active than we had any reason to assume based on our knowledge prior to Voyager.[26] The Voyagers played a major role in elevating planetary studies once again to a major status within the astronomy community and in the broad public consciousness.

There was, and is, more to be done. The original Grand Tour had included a planned encounter with the Solar System's most remote and enigmatic planet, Pluto. The Voyagers could not be programmed to accomplish that mission, and there were, at the beginning of the twenty-first century, no firm plans to visit this distant world and its satellite Charon. But Voyager has inspired spin-off exploratory missions, and some of those in turn are finding new discoveries and provoking new issues and unanswered questions. Meanwhile, the twin Voyagers, launched in 1977, continue their interstellar mission and hold the promise of yet further revelations about our Solar System and the universe.

In the year 2000 Voyager 1 was located more than 7 billion miles from the Sun (and just about that far from Earth), and Voyager 2 was almost 6 billion miles out. It took more than twenty-one hours for a radio signal traveling at the speed of light to make the round trip from Earth to Voyager 1 and back to Earth again; for Voyager 2 the two-way journey took almost seventeen hours. Estimates of their continued life-spans vary, but even conservative appraisals are that the Voyagers will soldier on until the year 2020 or even 2030, when their consumables, the hydrazine and the electric power from the RTGs, will no longer support operation of their science instruments. At that time Voyager 1 will be 150 times farther from the Sun than is Earth—14 billion miles or 20 billion kilometers distant—and Voyager 2 will be a few billion miles behind. They will then continue their journey through interstellar space. Neither spacecraft will pass close to any known nearby stars, but inevitably they will be out there among those stars—emissaries to *whatever* and, perhaps, *whoever* may lie beyond the limits of our own planetary system.[27]

APPENDIX

VOYAGER PROJECT MANAGERS

Manager	Years	Mission Phase
Harris Schurmeier	1970–76	Grand Tour/MJS '77 (inception/development)
John Casani	1976–78	Voyager/Launch
Robert J. Parks	1978–79	Cruise/Jupiter encounter
Raymond L. Heacock	1979–81	Jupiter/Cruise/Saturn encounters
Esker K. Davis	1981–82	Saturn encounters
Richard P. Laeser	1982–87	Cruise/Uranus encounters*
Norman R. Haynes	1987–89	Neptune encounter
George Textor	1989–99	Interstellar mission
Ed B. Massey	1999–	Interstellar mission

*Also head of the NASA Space Station Support Office, Reston, Virginia, 1986–87.

NOTES

1. FROM EARTH TO THE EDGE OF THE UNIVERSE

1. Jet Propulsion Laboratory, Media Relations Office, Pasadena California (here-after cited as JPL, Media Relations Office), "Voyager 1 Now Most Distant Human-Made Object in Space," February 13, 1998, and "Two Voyager Spacecraft Still Going Strong After 20 Years," September 2, 1997, both on line at www.jpl.nasa.gov.

2. JPL, Media Relations Office, "Fact Sheet: The Voyager Planetary Mission," March 8, 1998; Curtis Peebles, "The Original Voyager: A Mission Not Flown," *Journal of the British Interplanetary Society* 35 (1982): 9–15.

3. Charles Emile Kohlhase Jr., interview with Dethloff and Schorn, La Canada, Calif., January 25, 1999.

4. Ibid.

5. Ibid.

6. JPL, Media Relations Office, "*Voyager*'s Interstellar Mission," March 8, 1998.

7. *Star Trek: The Motion Picture,* Paramount Pictures, 1979.

8. JPL, Media Relations Office, "GEE-WHIZ Stuff about the Voyager Project," March 8, 1998; *Star Wars,* Lucas Films, 1977, and additional episodes; Arthur C. Clark, screenplay, *2001: Space Odyssey,* 1968.

9. Homer E. Newell, *Beyond the Atmosphere: Early Years of Space Science,* NASA SP-4211, Washington, D.C., 16–32; and see Walter A. McDougall, . . . *the Heavens and the Earth: A Political History of the Space Age* (New York: Basic Books, 1985), 18–62.

10. Newell, *Beyond the Atmosphere,* 16–32; McDougall, . . . *the Heavens and the*

239

Earth; Jane Van Nimmen and Leonard C. Bruno, with Robert L. Rosholt, *NASA Historical Data Book,* vol. 1, *NASA Resources 1958–1968,* NASA SP-4012, Washington, D.C., 1988, 519–20.

11. Newell, *Beyond the Atmosphere,* 16–32; McDougall, . . . *the Heavens and the Earth;* Jane Van Nimmen and Leonard C. Bruno, with Robert L. Rosholt, *NASA Historical Data Book,* vol. 1, *NASA Resources 1958–1968,* NASA SP-4012, Washington, D.C., 1988, 519–20.

12. Clayton R. Koppes, *JPL and the American Space Program* (New Haven: Yale University Press, 1982), 1–4.

13. Frank J. Malina, interview with R. Cargill Hall, October 29, 1968, JPL Center Papers, NASA History Office, Washington, D.C.

14. Ibid.; Koppes, *JPL,* 18–24.

15. WAC was to honor the Women's Auxiliary Corps, but it also was used as an acronym to signify "without attitude control." JPL engineers and administrators believed that the WAC Corporal had indeed exceeded 40 miles, but analysis of the radar data over several years indicated that false readings had been obtained. See John Bluth, NASA, Jet Propulsion Laboratory Archives, Pasadena, Calif. (hereafter cited as JPL Archives), Notes on the WAC Corporal (draft), November 1996.

16. Ibid. While 99 percent of the gaseous envelope surrounding Earth is within 50 miles of Earth's surface, the ionosphere, ranging from 50 to 400 miles beyond the surface contains high concentrations of electrically charged particles (ions). The final atmospheric layer is the exosphere, which thins into the vacuum of space. There is no question that a Bumper WAC flight, that is, a WAC (B) Corporal rocket mounted on a German V-2 rocket, attained an altitude of 244 miles on February 24, 1949, and thus qualified more fully than its antecedent for the distinction of having surmounted Earth's atmosphere. The Corporal became the ancestor of later rockets that were used as part of the Voyager interplanetary explorer launch systems.

17. "Papers Relating to the Armed Service Preparedness Investigating Subcommittee, Other People's Statements on Satellites," memorandum, Senate Papers, Box 356, Lyndon B. Johnson Library, Austin, Texas; and see Asif Siddiqi, *Challenge to Apollo: The Soviet Race to the Moon, 1945–1974,* NASA SP-2000-4408, Washington, D.C., 2000. See also Paul Dickson, *Sputnik: The Shock of the Century* (New York: Walker, 2001).

18. Senate Committee on Armed Services, Preparedness Investigating Subcommittee, *Inquiry into Satellite and Missile Programs, Hearings before the Preparedness Investigating Subcommittee of the Committee on Armed Services,* 85th Cong., 1st and 2d sess., part 1, pp. 1–2 (hereafter cited as Preparedness Subcommittee Hearings).

19. Craig B. Waff, "The Road to the Deep Space Network," *Spectrum,* April 1993, 50–57.

20. T. Keith Glennan, *The Birth of NASA: The Diary of T. Keith Glennan,* NASA SP-4501, edited by J. D. Hunley, with an introduction by Roger D. Launius, Washington, D.C., 1993, 12.

21. Ibid.; Linda Neuman Ezell, *NASA Historical Data Book,* vol. 2, *Programs and Projects, 1958–1968,* NASA SP-4012, Washington, D.C., 1988, 300–304. Note: Ramo-Woolridge Corporation later became Thompson-Ramo-Woolridge Corporation; Glennan, *Birth of NASA,* 12.

22. Waff, "Road to the Deep Space Network," 53; and see William R. Corliss, *A History of the Deep Space Network,* NASA CR-151915, Washington, D.C., May 1, 1976, 1–217.

23. Waff, "Road to the Deep Space Network"; and see Corliss, *History of the Deep Space Network,* 16–19; Douglas J. Mudgway, *Uplink-Downlink: A History of the Deep Space Network, 1957–1997,* NASA SP-2001-4227, Washington, D.C., 1–674.
24. Koppes, *JPL,* 96–100.
25. "Staff Report of the Select Committee on Astronautics and Space Exploration," *The Next Ten Years in Space* (Washington, D.C.: GPO, 1959), 15.
26. Ibid., 32.
27. Ibid.; Koppes, *JPL,* 92–93.
28. Koppes, *JPL,* 102–5.
29. Ibid., 92–93, 103–5.
30. Waff, "Road to the Deep Space Network," 50–57; Ezell, *NASA Historical Data Book* 2:302–9.
31. A Plan for Manned Lunar and Planetary Exploration, by Allan B. Hazard, JPL, November 1959, 3-323, Historical Collections, JPL Archives (hereafter cited as HC).
32. Ibid.
33. Ezell, *NASA Historical Data Book,* vol. 2, *Programs and Projects, 1958–1969,* 300–309.
34. Koppes, *JPL,* 117–19; Nicks, *Far Travelers,* 104–9.
35. Oran W. Nicks, interview with R. Cargill Hall, August 26, 1968, JPL Archives.
36. Ibid.
37. Koppes, *JPL,* 112.
38. Nicks, *Far Travelers,* 14–17; Later, in November 1962, adding to the confusion, a new and redesigned Pioneer program with interplanetary objectives received NASA approval, but was assigned to NASA's Ames Research Center.
39. Koppes, *JPL,* 109–12; Nicks, *Far Travelers,* 14–17; Ezell, *NASA Historical Data Book* 2:334.
40. Nicks, *Far Travelers,* 16–17.
41. Ibid., 17.
42. D. Schneiderman, Section Report No. 29-1, Spacecraft Design Criteria and Considerations; General Concepts, Spacecraft S-1, Jet Propulsion Laboratory, California Institute of Technology, Pasadena, California, February 1, 1960, 2 967, HC, JPL Archives.
43. Ibid; Schneiderman recognized that "the design of a space vehicle is constrained by a unique contradiction; it must function for the great majority of its lifetime in vast space and at zero gravity while during a brief moment at the launching it must withstand high acceleration vibrations and physical confinement." Weight was an essential element mandating a "lean and clean" design, but conversely, the basic design had to be sufficiently flexible and complex to accommodate the changes required by the unique character of each mission.
44. Ibid.
45. Ibid., 335; Nicks, *Far Travelers,* 1–5; Dan Schneiderman, interview with Robert Needell, July 17, 1982, JPL Archives.
46. Ezell, *NASA Historical Data Book* 2:334–38.
47. Ronald F. Draper, interview with Dethloff and Schorn, JPL, Pasadena, Calif., January 29, 1999. Voyager History Project Papers, NASA Historical Records Collections, NASA History Office, Washington, D.C. (hereafter cited as NASA HRC) and JPL Archives.
48. Ibid.

49. John Richard Casani, interview with Dethloff and Schorn, JPL, January 28, 1999.
50. Ibid.
51. Ibid.; "To Mars: The Odyssey of Mariner IV," JPL Technical Memorandum No. 33-229, NASA, 1965, i, 1–8.
52. Bruce C. Murray, Voyager Dialogue files, 3-481, HC, JPL Archives.
53. California Institute of Technology, "Suggestions for Martian Exploration Following Mariner IV," February 23, 1965, 3-481, HC, JPL Archives; Curtis Peebles, "The Original Voyager: A Mission Not Flown," *Journal of the British Interplanetary Society* 35 (1982): 9–15; Homer E. Newell to Dr. Lee A. DuBridge, August 16, 1965, 3-481, HC, JPL Archives; *NASA News* Release No. 65-242, July 20, 1965, NASA HRC.
54. *NASA Pocket Statistics,* 1996, B-92 to B-97.
55. Dedication of the Goldstone Antenna, 7-11a, 7-11b, HC, JPL Archives; R. J. Parks, interoffice memorandum, February 4, 1966, 6-215, 6-223, HC, JPL Archives; Garbarini to H. H. Haglund, and Garbarini to R. J. Parks, June 6, 1966, 6-221, HC, JPL Archives; Mudgway, *Uplink-Downlink,* 166–93.
56. Peebles, "Original Voyager," 12–13.
57. *NASA Pocket Statistics,* 1996 ed., B-96 to B-99; Peebles, "Original Voyager," 14–16; *Space Daily,* September 6, 1967, 18; Summary of the Voyager Program, NASA, Office of Space Science and Applications, January 1967, 3-283, HC, JPL Archives; Koppes, *JPL,* 173–74; JPL, Section Chief Briefing, 12/17/65, 3-455, HC, JPL Archives.
58. John E. Naugle, memorandum, May 15, 1968, Voyager Files, NASA HRC.

2. A GRAND TOUR OF THE OUTER PLANETS

1. Jet Propulsion Laboratory, *1967 Annual Report* (Pasadena, Calif.: JPL, 1967), 11.
2. See Forest Ray Moulton, *An Introduction to Celestial Mechanics,* 2d rev. ed. (New York: Macmillan, 1914), 277–319, is a chapter devoted entirely to special solutions of the restricted three-body problem. A historical sketch and bibliography at the end of the chapter are especially informative.
3. Donald K. Yeomans, *Comets: A Chronological History of Observation, Science, Myth, and Folklore* (New York: John Wiley, 1991), 157–60. This is the best recent book about comets, bar none.
4. Rex Ridenoure to Dethloff, editorial comment on preliminary draft of chapter 6, May 10, 1999, Voyager History Project Papers, JPL Archives.
5. Robert A. Heinlein, *The Rolling Stones* (New York: Charles Scribner's Sons, 1952). One wonders if the much later rock group of the same name ever heard of this novel.
6. Andrew J. Butrica, "Voyager: The Grand Tour of Big Science," in *From Engineering Science to Big Science: The NACA and NASA Collier Trophy Research Project Winners,* ed. Pamela E. Mack, NASA SP-4219 (Washington, D.C.: NASA, 1998), 254.
7. Victor C. Clarke Jr. to Norriss S. Hetherington, October 16, 1972, 5-633, HC, JPL Archives. As one example of a discrepancy concerning this matter, Richard A. Dowling, William J. Kosmann, Michael A. Minovitch, and Rex W. Ridenoure, "Origin of Gravity-Propelled Interplanetary Space Travel" (paper presented at the 41st Congress of the International Astronautical Federation, October 6–12, 1990, Dresden, GDR), state that Minovitch was initially assigned the problem of calculating one-way trajectories from Earth to another planet. Off-print courtesy Michael A. Minovitch.
8. Victor C. Clarke Jr. to Norriss S. Hetherington, July 22, 1974, provided the au-

thors by Michael A. Minovitch; Elliott Cutting to Norriss Hetherington, October 25, 1972, 5-623, HC, JPL Archives. The reader may be interested to know that Hetherington was a historian at the University of Kansas who wrote a history of gravitational assist. When he circulated an early draft of his manuscript he received some frank and detailed criticism. His manuscript, along with responses to it, provide the basic information for this section.

9. Ridenoure to Dethloff, editorial comment on preliminary draft of chapter 6, May 10, 1999.

10. Voyager Mission Planning Office Staff, Charles Kohlhase, ed., *The Voyager Neptune Travel Guide* (Pasadena, Calif.: JPL, 1989), 103–5, gives a succinct description of early work on developing the gravity assist technique into a practical means of attaining planetary targets, as well as its earliest applications. The book itself was a press guide for the encounter of Voyager 2 with Neptune.

11. Gary A. Flandro, "Discovery of the Grand Tour *Voyager* Mission Profile," in *Planets Beyond: Discovering the Outer Solar System,* ed. Mark Littman (New York: John Wiley, 1988), 95–98.

12. Rex Ridenoure notes that Michael Minovitch lacked access to precision ephemerides for the post-1980 period for Saturn and beyond and thus could not establish the detailed trajectories determined by Flandro. Ridenoure to Dethloff, editorial comment on preliminary draft of chapter 6, May 10, 1999.

13. Ibid. Ridenoure indicates that Michael Minovitch had previously noted the Grand Tour planetary profile in his personal journals in 1962. Flandro, in 1965, identified the alignment as having "mission significance." Flandro, "Discovery," 95–98.

14. G. A. Flandro, "Fast Reconnaissance Missions to the Outer Solar System Utilizing Energy Derived from the Gravitational Field of Jupiter," *Astronautica Acta* 12 (1966): 329.

15. H. J. Stewart to George M. Henry, February 23, 1970. Henry was writing an article on planetary exploration and on February 13, 1970, wrote to Stewart asking him to "pin down" the very first mentions of "gravity assist" (swingby, any planet), "serial swingby, outer planets, late 1970's," and the "expression Grand Tour." Incidentally, Stewart wrote that the expression "Grand Tour" came out of Minovitch's work, although *Time* used it as early as 1963 (though with no capital letters).

16. To give just one example: the *Denver Post,* November 27, 1966, ran a story headlined "'Billiards' Approach in Mind for Neptune," complete with a diagram of the proposed trajectory. The interested reader can examine the files of a local newspaper at the time and will probably find a similar item.

17. *Time,* August 23, 1968, 56.

18. John R. Strand to Dethloff and Schorn, memorandum and letter, August 9, 2000, Voyager History Project Papers, JPL Archives.

19. James E. Long, "To the Planets," *Astronautics & Aeronautics,* June 1969, 32. Our discussion here is basically Long's, for his account of these early studies concerning missions to the outer planets is much to be preferred to other, more detailed (and usually less accessible) reports, briefings, and the like. If the reader wants to look up only a single reference on the subject, this is the one. And see the unpublished 1972 Jupiter Flyby Mission Study, ASD 760-22, 15 July 1968, 8-248, HC, JPL Archives.

20. Long, "To the Planets."

21. Ibid.

22. Ibid.

23. Ibid.

24. Various authors, *Astronautics & Aeronautics,* special issue, September 1970, 35–95. Of special interest is the lead-off summary article by Rob R. McDonald and William S. Shipley. Among the JPL documents relating to TOPS is "Thermoelectric Outer Planets Spacecraft: Industry Briefing," September 21, 1971, 10-5, HC, JPL Archives.
25. Bruce C. Murray to William Fowler, interoffice memorandum, December 14, 1971, 11-55, HC, JPL Archives; "Outer Planets Grand Tours Science Pre-Proposal Briefing," November 17, 1970, 10-4, HC, JPL Archives.

3. A FUNNY THING HAPPENED ON THE WAY TO NEPTUNE

1. Murray to Fowler, interoffice memorandum, December 14, 1971, 11-55, HC JPL Archives.
2. Bruce C. Murray, U.S. Planetary Decisions in 1968—A Test of National Judgment, White House/President's Scientific Advisory Committee Files, NASA HRC.
3. Harris M. Schurmeier, "Planetary Exploration: Earth's New Horizon," *Journal of Spacecraft and Rockets* 12, no. 7 (July 1975): 385–405.
4. Flandro, "Fast Reconnaissance"; Long, "To the Planets," 32–47.
5. JPL, California Institute of Technology, Office of Public Information, September 11, 1969, 8-263, HC, JPL Archives; William E. Blundell, "Jet Propulsion Lab Probes Other Planets Seeking Signs of Life," *Wall Street Journal,* February 21, 1969; James H. Wilson, *Two Over Mars, Mariner VI and Mariner VII, February to August 1969,* NASA EP-90 (Pasadena, Calif.: NASA-JPL/California Institute of Technology, n.d.), 1–37; *Mariner-Mars 1969: A Preliminary Report,* NASA SP-225 (Washington, D.C.: NASA, 1969), 1–26.
6. Blundell, "Jet Propulsion Lab."
7. *The Next Decade in Space: A report of the Space Science and Technology Panel of the President's Science Advisory Committee* (Washington, D.C.: Executive Office of the President, Office of Science and Technology, March 1970), i–ii, i–63.
8. "Jet Propulsion Laboratory, Outer Planets Grand Tour, Presentation to NASA SL, 29 June 1970," 10-3, 48, HC, JPL Archives.
9. JPL Five-Year Plan, April 1, 1971, 3-471, HC, JPL Archives.
10. Ibid.
11. "NASA Hopes to Fly Grand Tour Missions for $856 Million," *Space Daily,* June 28, 1971, 298; "Grand Tour Science Support Requests for Mission Definition (Summary)," memorandum, June 3, 1971, in Grand Tour Project Papers, JPL Archives.
12. "Outer Planets Grand Tour, Presentation to NASA SL," 10-3.
13. Ibid.
14. Draper interview.
15. Ibid.; and see "TOPS—Outer Planets Spacecraft," *Astronautics & Aeronautics* 8, no. 9 (September 1970): 35–95.
16. Draper interview; "TOPS—Outer Planets Spacecraft," 35–95.
17. "Outer Planets Grand Tour, Presentation to NASA SL," 10-3; "TOPS—Outer Planets Spacecraft," 35–95.
18. Ibid.
19. Raymond L. Heacock, interview with Dethloff and Schorn, JPL, Pasadena, Calif., February 4, 1999.
20. *NASA News* Release No. 71-56, April 4, 1971.
21. Associate Administrator for Space Science and Applications to Assistant Admin-

istrator, Office of Administrator, draft memorandum, March 26, 1971, NASA HRC.

22. Grand Tour Trip Times with Nerva, memorandum, May 26, 1971, NASA HRC.
23. NASA, *Space Daily,* June 28, 1971, 298; "Grand Tour Science Support Requests for Mission Definition (Summary)."
24. NASA, *Space Business Daily,* June 30, 1971, 317; *Space Business Daily,* July 23, 1971, 100.
25. Newell, *Beyond the Atmosphere,* 289; McDougall, . . . *the Heavens and the Earth,* 168; NASA, *Space Business Daily,* July 23, 1971, 100.
26. *Space Business Daily,* July 23, 1971.
27. *Space Business Daily,* September 23, 1971, 107–8; *Washington Evening Star,* January 6, 1971. In the perspective of Voyager discoveries and a decade of results from the Hubble Space Telescope (which has a primary mirror 96 inches in diameter), the idea that a 45-inch space telescope could produce more "good science" than the Grand Tour was palpably wrong.
28. William A. Fowler, Chairman, PSC/SPAC to Dr. Homer E. Newell, October 1, 1971, NASA HRC. Scientists' aversion to imagery had to do in part with the perception that anything easily understandable was not "good science." In addition, the committee's strong emphasis on the chemical compositions of the major planets, and its view that the smaller, airless bodies of the Solar System were scientifically uninteresting was disproved by the Voyager observations of Io and Europa, to mention just two examples.
29. A Staff Meeting Summary, September 28, 1971, NASA HRC.
30. Linda Neumann Ezell, *NASA Historical Data Book,* vol. 3, *Programs and Projects, 1969–1978,* NASA SP-4012 (Washington, D.C.: NASA, 1988), 213–18; SV/J.B. Mahon to Lewis Research Center (Attn: S. C. Himmel), memorandum, August 11, 1971, NASA HRC.
31. NASA, *Space Business Daily,* October 21, 1971, 182.
32. Henry C. Dethloff, "The Space Shuttle's First Flight: STS-1," in Mack, *From Engineering Science to Big Science,* 277–97.
33. Ibid.
34. *Space Business Daily,* December 8, 1971, 169.
35. *Space Business Daily,* December 14, 1971, 192.
36. "Impact of Deferring Grand Tour Launch," memorandum, R. S. Kraemer, December 15, 1971.
37. James C. Fletcher to Caspar W. Weinberger, December 22, 1971, NASA HRC; *Los Angeles Times,* January 26, 1972; Robert S. Kraemer, telephone interview with Dethloff, Annapolis, Md., April 28, 1999.
38. *Los Angeles Times,* January 26, 1971.
39. *Space Business Daily,* February 24, 1972.
40. Murray to Fowler, interoffice memorandum.
41. Ibid.
42. Ibid.

4. METAMORPHOSIS

1. "Who Killed Grand Tour"; Heacock interview.
2. Schurmeier, "Planetary Exploration," 385–405.
3. Ibid.
4. Draper interview.

5. Ibid.
6. Ibid.; John Richard Casani, interview with Dethloff, Canyon Lake, Calif., February 3, 1999.
7. Ibid.
8. Ibid.
9. Kraemer interview.
10. Ibid.; "NASA Proposes Jupiter-Saturn Mission," *NASA News* Release No. 72-42, February 24, 1972, NASA HRC.
11. Ibid; note that Naugle's cost estimate for Congress on MJS was considerably higher than the $250 million noted by Heacock; David W. Swift, *Voyager Tales: Personal Views of the Grand Tour* (Reston, Va.: American Institute of Aeronautics and Astronautics, 1997), 103–12.
12. Ibid.
13. Draper interview; Heacock interview; Casani interview, January 28, 1999.
14. Interview with Harris (Bud) Schurmeier in Swift, *Voyager Tales,* 107–8.
15. Schurmeier, "Planetary Exploration," 385.
16. John Naugle to William H. Pickering, April 12, 1972; Associate Administrator for Space Science to Deputy to AD Deputy Administrator, memorandum, October 13, 1972, p. 36, fol. 29, JPL Archives; J. K. Davies, "A Brief History of the Voyager Project," *Spaceflight* 23 (February 2, 1981): 35–41; and see Raymond L. Heacock, "The Voyager Spacecraft," James Watt International Gold Medal Lecture, Institute of Mechanical Engineers, *Proceedings of the Institution of Mechanical Engineers* 194, no. 28 (1980): 211–24.
17. Interview with William S. Shipley in Swift, *Voyager Tales,* 224–37.
18. Draper interview.
19. Ibid.; interview with Shipley in Swift, *Voyager Tales,* 226.
20. "Mariner Jupiter/Saturn 1977, Space Craft Description, July 12, 1972," Revised December 13, 1972; Revised April 25, 1974, Flight Programs, 44-84, JPL Archives.
21. Ibid.
22. Ibid.
23. Interview with Bourke in Swift, *Voyager Tales,* 76–79.
24. Kohlhase interview; interview with Charles Kohlhase in Swift, *Voyager Tales,* 83–100.
25. Kohlhase interview.
26. Ibid.
27. Interview with Thomas Gavin in Swift, *Voyager Tales,* 250–55.
28. Interview with Richard (Dick) Laeser in Swift, *Voyager Tales,* 173–80.
29. Ibid.
30. Edward Carroll Stone Jr., interview with Dethloff, JPL, Pasadena, Calif., February 4, 1999; interview with Edward C. Stone in Swift, *Voyager Tales,* 37–57.
31. Stone, interview with Dethloff, February 4, 1999.
32. "Grand Tour Scientists," *NASA News* Release No. 71-56; "Scientists Selected for Jupiter/Saturn Missions," *NASA News* Release No. 72-239, December 10, 1972, NASA HRC.
33. Ibid.
34. Ibid.
35. Stone, interview with Dethloff, February 4, 1999.
36. Kohlhase interview.
37. Ibid.
38. Ibid.

39. NASA Program Reductions, *NASA News* Release No. 73-3, January 5, 1973, NASA HRC; *Washington Evening Star and Daily News,* January 6, 1973; *Washington Post,* January 6, 1973.

5. MARINER JUPITER-SATURN '77

1. Jet Propulsion Laboratory, *1973–1974 Annual Report* (Pasadena, Calif.: JPL, 1974), 9; Heacock interview; Heacock, "Voyager Spacecraft," 211–24.
2. Ibid.
3. *NASA News* Release No. 73-3.
4. Heacock interview.
5. Heacock, "Voyager Spacecraft," 216.
6. Ibid., 221.
7. Ibid., 221–22; Heacock interview.
8. Ibid.
9. Heacock, "Voyager Spacecraft," 218–19; Heacock interview.
10. Heacock interview; emphasis added.
11. Ibid.
12. Ibid., 222–23; Ezell, *NASA Historical Data Book* 3:21–26, 38–42.
13. Ibid.
14. Kohlhase interview.
15. Ibid.
16. Ibid.
17. Heacock interview.
18. Ibid.; Heacock, "Voyager Spacecraft," 224.
19. Ezell, *NASA Historical Data Book* 3:129–31; Voyager 1 and 2 Press Kit, *NASA News* Release No. 77-136, August 4, 1977, "Voyager Science," press handout, 108.
20. Heacock interview.
21. Draper interview.
22. Ibid.
23. Ibid.
24. Ibid.
25. Harris M. Schurmeier to MJS 77 Review Board, November 6, 1974, Flight Collections, p. 36, fol. 54, JPL Archives (hereafter cited as FC).
26. Ibid.
27. Ibid.
28. Mariner Jupiter/Saturn 1977, Mission & Systems Design Review, Concern/Action Control Sheet, 16/17 October 1974, p. 36, fol. 29, FC, JPL Archives.
29. Ronald A. Schorn, *Planetary Astronomy: From Ancient Times to the Third Millennium* (College Station: Texas A&M University Press, 1998), 202, 212, 227–28, 361.
30. "Voyager Will Carry 'Earth Sounds' Record," *NASA News* Release No. 77-159, August 1, 1977; and see Carl Sagan, *Murmurs of Earth: The Voyager Interstellar Record* (New York: Ballantine Books, 1983).
31. Voyager's Interstellar Outreach Program, Voyager Home Page, vraptor.jpl.nasa.gov, March 8, 1998.
32. Ibid.
33. *NASA News* Release No. 77-159.
34. Mariner Jupiter Saturn 1977 Mission Plan, 2 December 1974, p. 90-061, fol. 166, FC, JPL Archives.

35. Harris M. Schurmeier to MJS 77 Review Board, memorandum MJS-HMS-74-46, Project Actions in Response to Review Concerns, p. 36, fol. 54, FC, JPL Archives.
36. *Defense/Space Daily,* April 4, 1975, 198.
37. Information from ERDA, No. 75-79, May 21, 1975, JPL Center Papers, NASA HRC.

6. VOYAGER

1. "Harris M. (Bud) Schurmeier," in *Voyager Tales: Personal Views of the Grand Tour,* ed. David W. Swift (Reston, Va.: American Institute of Aeronautics and Astronautics, 1997), 108–9.
2. Dethloff, "Space Shuttle's First Flight," 277–97.
3. Ibid.; and see Dethloff, "The Space Shuttle's First Flight," 277–97.
4. Swift, *Voyager Tales,* 108–9.
5. Ezell *NASA Historical Data Book* 2:197–99; Ezell, *NASA Historical Data Book* 3:127–30, 235–39.
6. Koppes, *JPL,* 241–42.
7. Swift, *Voyager Tales,* 103, 108–9.
8. Casani interview, January 28, 1999.
9. Ibid.
10. Ibid.
11. Ibid.
12. Ibid.
13. Ibid.
14. Ibid.
15. Ibid.
16. Ibid.
17. Ibid.
18. Ibid.
19. Ibid.
20. NASA, Project Approval Document, Code Number 802, February 10, 1976, JPL Center Papers, NASA HRC.
21. Ibid.
22. John Noble Wilford, "Uranus Mission Approved by NASA," *New York Times,* March 1, 1976.
23. *NASA News* Release No. 77-41, March 7, 1977.
24. Joseph P. Allen to Honorable Jack Brooks, August 2, 1977, JPL Center Papers, NASA HRC.
25. Ibid.
26. Ibid.
27. Heacock interview.
28. *NASA News* Release No. 77-136.
29. Davies, "Brief History of the Voyager Project," 35–39; Heacock interview.
30. Ibid.; "Voyager Switch Avoids Delay in Launch," *Aviation Week & Space Technology,* August 19, 1977, 20–21.
31. *NASA News* Release No. 77-136.
32. Ibid.
33. Ibid.
34. Ibid.

35. Swift, Interview with Richard Laeser, 175–80; interview with Robert (Bob) Parks, in Swift, *Voyager Tales,* 121–31.
36. Voyager, Mission Operations Management Plan, December 1, 1977, RG 44, fol. 152, JPL Archives.
37. Voyager, Mission Status Bulletin, No. 4, August 25, 1977, JPL Center Papers, NASA HRC; J. K. Davies, "A Brief History of the Voyager Project, Part 2," *Spaceflight* 23, no. 3 (March 1981): 69–74; *New York Times,* September 4, 1977.
38. Phillip H. Abelson, "The Voyager Missions," *Science* (9 September 1977).

7. VOYAGER SCIENCE

1. Edward Carroll Stone Jr., interview with Dethloff and Schorn, JPL, Pasadena, Calif., January 28, 1999, Voyager/Intaglio Papers, NASA HRC and JPL Archives.
2. Ibid.
3. Ibid.; *NASA News* Release No. 77-136, 1–3.
4. *NASA News* Release No. 77-136, 68–69.
5. Stone interview, January 28, 1999.
6. Ibid.
7. Ibid.
8. *NASA News* Release No. 77-136, 68–69.
9. Ibid., 6–7.
10. Ibid., 108.
11. Ibid., 108–11.
12. Voyager Mission Operations Management Plan, 618-77, Flight Records 44-152, JPL Archives.
13. Stone interview, January 28, 1999.
14. Ibid.; *NASA News* Release No. 77-136, 86–90.
15. Stone interview, January 28, 1999.
16. These recommendations involved giving ratings to particular proposals, essentially rating each one on a scale that ranged from "must fly" to "forget it," along with specific reasons for that rating. However, the written record leaves a lot out. One of us (R. A. S.) served on several selection boards (though not on the Voyager panels) and can attest that there was a great deal of discussion about various proposals that was never written down. At the same time, there was not a lot of special pleading, for the panels were large enough that anyone attempting such a tactic was likely to have been voted down. Regarding Voyager in particular, both our formal interviews and informal conversations failed to uncover anyone that was really sore about the selection process.
17. Ibid.
18. Ibid.
19. Ibid.
20. Bradford A. Smith, interview with Schorn, Austin, Tex., January 8, 1999, NASA HRC.
21. Ibid.
22. Ibid.
23. Garry E. Hunt, telephone interview with Dethloff, London, England, April 6, 1999.
24. See Butrica, "Voyager," xi–xii, 251–76.
25. Voyager Mission Operations Management Plan, 618-77.

26. Ibid.
27. Stone interview, January 28, 1999.
28. Ibid.
29. Ibid.
30. *NASA News* Release No. 77-136, 68.
31. Ibid.

8. LAUNCH

1. Leonard (Pete) Piasecki, a senior solid-propellant rocket expert at JPL, puts this succinctly. "Rocket engines," he maintains, "are alive."
2. The trade magazine *Aviation Week* contains literally hundreds of articles, news notes, and items relating to Titan launch vehicles dating from the 1960s to the present. An informative though difficult-to-find publication is *Titan III for Europe,* published by the Martin Marietta Corporation in June 1969, which gives the estimated performances of a wide range of projected Titan III versions, including some that never flew. As an aside, the Titans' diameters, along with those of some other U.S. rockets, were (and in some cases still are) related to dimensions such as height clearances on Interstate highways, sizes of nuclear warheads, and girths of reconnaissance and early warning satellites. Of course there are few references to the subject in the open, contemporary literature.
3. *JPL and Space Solid Rockets: A Historical Perspective of Solid Propellant Motors Developed or Used by JPL (1958–1988),* JPL D-2396 (Pasadena, Calif.: JPL, n.d).
4. Details of the Voyager launch vehicles may be found in many sources. One example is *Voyager Mission Status Bulletin No. 4,* JPL Center Papers, NASA HRC. The risky nature of planetary missions was demonstrated long after the Voyager encounters with the outer planets when the Mars Climate Observer in 1990, and the Mars Polar Lander in 1999, simply "went off the air" in the immediate vicinity of the red planet. Perhaps the saddest record of all was set by the Soviet Union which, despite literally dozens of attempts, never did have a successful Mars probe. A good overall summary of the planned Voyager missions is in *NASA News* Release No. 77-136.
5. J. K. Davies wrote a concise, informative series of articles for the journal *Spaceflight* under the running title of "A Brief History of the Voyager Project," with emphasis on engineering and operations rather than scientific results. The first installment began on page 35 of the February 2, 1981, issue and covered the history of the project and its antecedents up to the time of the first launch; the second, page 71 of the March 3, 1981, issue, continued the story from the two launches to the beginning of Voyager 1's encounter with Jupiter, while a third, on page 129 of the May 5, 1981, issue covered Voyager 1's passage through the Jovian system.
6. *Aviation Week & Space Technology,* September 5, 1977, 19.
7. *Aviation Week & Space Technology,* September 12, 1977, 20. From the late 1970s on, this journal carried literally hundreds of items concerning Voyager, many if not most dealing with nonglamorous events not connected with planetary encounters and unreported by newspapers, radio, or television.
8. Ray Cline, Associate Administrator for Management Operations at NASA Headquarters, to Distribution, cover letter, JPL Center Papers, NASA HRC. The copy that we have seen has the date "Jul 9 1979" stamped on it, but there is no indication of whether that refers to the date of sending or receiving. However, the cover letter does note that the period covered by the report extended from October 1,

1977, to September 30, 1979. Conflicts between JPL and NASA Headquarters have erupted every so often ever since President Eisenhower transferred the laboratory from the U.S. Army to NASA. The dustup covered here was rather mild compared to those such as the one that followed the failures of the first six Ranger lunar probes. In that case Pickering was almost fired and had to accept a retired army general, Alvin R. Leudecke, as deputy director to handle JPL's day-to-day affairs. At the time one of us (R. A. S.) was a JPL employee detailed to NASA Headquarters and vividly remembers the situation, which involved a full-blown congressional investigation. For details see Koppes, *JPL*. In a curious postscript, both authors later knew General Leudecke when he was acting president at Texas A&M University and we were faculty members.

9. Cline to Distribution, July 9, 1979.
10. Ibid.
11. Ibid.
12. A few of the many Voyager test scripts include 618-515, Voyager Mission Operations Test Plan, vol. 3-B, Flight Team Test and Training Jupiter Encounter, August 1978, JPL 93, Box 2, Folder 22; 618-790, Voyager JST Near Encounter Test Script, Test Date 12 December, 1978, signed off 12 December, 1978, JPL 93, Box 2, Folder 25; and 618-791, Voyager Project CCS Abort Test Script, signed off on November 17, 1978, JPL 93, Box 2, Folder 24, all in JPL Archives.
13. Ellis Miner to Henry C. Dethloff, September 8, 2000, NASA Historical Collections.
14. The JPL internal document is 618-51, Rev. C, Mariner Jupiter/Saturn 1977, Mission and Science Requirements Document, vol. 1, Mission Requirements, August 1, 1976, 2–1. The NASA press kit is Release No. 77-136, dated August 4, 1977, while the *Science* editorial by Philip H. Abelson is in the September 9, 1977, issue.
15. Carl Sagan got into hot water for making that statement, but one of us (R. A. S.) was a graduate student at Yerkes when Kuiper was director, and Carl was right.
16. Robert H. Baker, *Astronomy*, 6th ed. (Princeton, N.J.: D. Van Nostrand, 1955), 205–21. Tables of planetary and satellite properties are on 223–24.
17. Robert H. Baker, *Astronomy*, 7th ed. (Princeton, N.J.: D. Van Nostrand, 1959), 217–31.
18. Henry Norris Russell, Raymond Smith Dugan, and John Quincy Stewart, *Astronomy: A Revision of Young's Manual of Astronomy*, vol. 1 (Boston: Ginn and Company, 1945), 361–403. The junior author of this work bought his new (not used) copy in 1964. Pioneer planetary scientists such as Michael Belton, Dale Cruikshank, Donald Hunten, Raymond Newburn, Tobias Owen, Carl Sagan, Hyron Spinrad, and Louise Young, to name a few, all used this book.
19. N. T. Bobrovnikoff in *Reviews of Modern Physics* 16 (1944): 271.
20. Gerard P. Kuiper, ed., *The Sun* (Chicago: University of Chicago Press, 1953); Gerard P. Kuiper, ed., *The Earth as a Planet* (Chicago: University of Chicago Press, 1958); Gerard P. Kuiper and Barbara M. Middlehurst, eds., *Planets and Satellites* (Chicago: University of Chicago Press, 1961); and Barbara M. Middlehurst and Gerard P. Kuiper, eds., *The Moon, Meteorites, and Comets* (Chicago: University of Chicago Press, 1963).
21. Schorn, *Planetary Astronomy*. Chapters 6 to 11 survey the course of Solar System research from the 1930s up to the time of the Voyager missions.
22. *The Andromeda Strain* was a 1971 movie directed by Robert Wise and based on the novel by Michael Crichton in which a deadly virus from another galaxy threatens to destroy all life on Earth. By no coincidence at all, some of the scenes were filmed at JPL!

23. For the plan, see, for example, Mariner Jupiter/Saturn 1977 Planetary Quarantine Plan, PD-618-114, August 1, 1976. For the successful result, see Voyager Planetary Quarantine Analysis Part II: Post-Launch Analysis, February 1, 1978, JPL 93, Box 2, folders 20, 21, JPL Archives.
24. Mudgway, *Uplink-Downlink,* 166.

9. FIRST ENCOUNTERS

1. There are myriad citations to the scientific results of this encounter, but the single most convenient, easily obtainable, and most important source is a special issue (vol. 204) of *Science* for June 1, 1979, containing fifteen articles on pages 945–1008 and later reprinted as a separate booklet by the American Association for the Advancement of Science as Publication 79-6. In general, each paper discusses the results of a particular experiment, for example, visible-light imaging, infrared observations, ultraviolet observations, and the like. This convenient collection covered the entire range of scientific studies and, although various authors were careful to state that the results described therein were merely preliminary, soon became the benchmark with which later work was compared. Special issues of *Science* summarizing first results became a Voyager trademark and eventually appeared after all six planetary encounters. Then, too, generally a year or two after each planetary encounter appeared a hefty special edition of the *Journal of Geophysical Research;* volume 85, pages 8123–41, came out in 1981 and presented results from both the Voyager 1 and Voyager 2 Jupiter encounters.

   A special issue (vol. 280) of *Nature* dated August 30, 1979, which contained twenty-seven articles on Jupiter and Io occupies pages 725–806. However, this was a collection of studies on specific topics such as sulphur flows on Io and Jupiter's magnetic tail. Later there appeared a magnificent volume titled *Time Variable Phenomena in the Jovian System,* NASA SP-494, Washington, D.C., 1989, edited by Michael J. S. Belton, Robert A. West, and Jürgen Rahe, which has the finest illustrations that we have seen concerning the encounters of both Voyagers 1 and 2 with the Jovian system.

   In this work we are basically telling things as they happened, rather than viewing with hindsight, which is always 20-20. Therefore, the special issues of *Science* will be the principal references on which we rely. In the interests of readability we have eschewed alike arranging results by instrument (television cameras, radio receivers, etc.), type of phenomenon (magnetic fields, surface features, and so forth), or in strict chronological order. Instead, we proceed in the main from one topic to another (the Galilean satellites, atmospheric motions on Jupiter, radio emissions, and the like. Because of this arrangement we have not burdened the text with individual references to specific items (such as the initial detection of active volcanoes on Io) to spare the reader from the more or less repetitive citations that would add up to many hundreds in this chapter alone. The context of our remarks should indicate clearly which instrument was involved and, as none of the articles in the special *Science* issue is overly long, the reader should have little trouble in finding the proper information. And see also *Science* 206 (1979): 925–96, *Geophysical Research Letters* 7 (1980): 1–68, and *Voyage to Jupiter,* NASA SP-439, Washington, D.C., 1980.
2. Every single interviewee we asked, engineer and scientist alike, offered "Io" as the one word answer to the question of what was the most unexpected and spectacular discovery of the entire Voyager program.

3. Stanton J. Peale, Patrick Cassen, and Ray Thomas Reynolds in *Science* 203 (1979): 97.
4. See, for example, Harold Masursky et al. in *Nature* 280 (1979): 725 and M. H. Carr et al. in ibid., 729.
5. *Physics Today,* February, 2000, 20.
6. Try breaking and then repairing a china plate, a glass, or—best of all—an ice cube, and note how difficult it is to hide all evidence of the break. Such analogies are in no sense exact, but they are illustrative. To make your own model of Europa, take a hard boiled egg, crack it gently all over, and dunk it for a bit in strong tea.
7. M. H. Acuña and N. F. Ness in *Journal of Geophysical Research* 81 (1976): 2917.
8. Bertrand M. Peek, *The Planet Jupiter* (London: Faber and Faber, 1958), 97.
9. Peek's book, which is referenced in the above note, gives a comprehensive review of observational results concerning the planet from the earliest days until the time of publication. The emphasis is on visual observations of cloud features, work which was primarily carried on by amateur astronomers, as many professionals believed that such investigations were beneath their dignity.
10. A good illustration of the "cumulonimbus" nature of these white spots is in G. E. Hunt and J.-P. Muller in *Nature* 280 (1979): 779. There is a vivid image of the "cauliflower-like" top of a rising convective cell in Bradford A. Smith et al., *Science,* June 1, 1979, 954 (fig. 4 in the article).
11. An easily available rendition of Jove forging and hurling thunderbolts is the Beethoven's "Pastoral Symphony" portion of Walt Disney's "Fantasia." An equally vivid case of mythological "prediction" involves dragons, which were epic staples long before dinosaurs were recognized; "Beowulf" is a good example.
12. The basic reference for the Voyager 2 flyby of Jupiter is the special November 23, 1979, issue of *Science.*
13. These are the numbers given on page 937 of the special November 23, 1979, issue of *Science.* Later references refer to "8 out of 9" because a lot of images had been looked at more closely in the interim.
14. Convenient lists of planetary satellites are found in many places, including most elementary astronomy texts. One listing is in J. Kelly Beatty, Brian O'Leary, and Andrew Chaikin, eds., *The New Solar System,* 2d ed. (Cambridge, Mass.; Sky Publishing, 1982), 220–21. Ellis D. Miner, "Revision of SFO Science and Mission Systems Handbook (PD 618-128)," a JPL interoffice memorandum dated March 15, 1990 (JPL 105, Box 2, Folder 30), lists the satellites of the outer planets on pages 18-2 and 18-3 and is in many other ways a gold mine of useful information.
15. Rex Ridenoure to Henry C. Dethloff, August 27, 2000.
16. Belton, West, and Rahe, *Time Variable Phenomena,* combines Voyager and ground based observations with theoretical studies of Io and Jupiter's atmosphere.
17. *Journal of Geophysical Research* 86 (1981): 8123–41 contains a comprehensive survey of the Voyager 1 and 2 encounters with Jupiter.
18. Ellis Miner, interview with Dethloff and Schorn, JPL, Pasadena, Calif., January 26, 1999, Intaglio/Voyager Papers, NASA HRC and JPL Archives.

10. SATURN'S SYSTEM

1. The limit of horizontal visibility at Earth's surface is about 300 kilometers (200 miles), which the reader can check without any special equipment. The southwestern United States is a good place to try such an experiment, but you have to be on a relatively high location and have a "clear shot" to another. At the limit,

distant objects will appear extremely faint, even if snow covered; viewing them against the sunrise or sunset horizon will help.

2. The standard reference on Saturn research through the 1950's is A. F. O'D. Alexander, *The Planet Saturn: A History of Observation, Theory, and Discovery* (London: Faber and Faber, 1952).

3. Stone interview, January 28, 1999.

4. Rex Ridenoure, interview with Dethloff, JPL, Pasadena, Calif., January 27, 1999, Voyager/Intaglio Papers, NASA HRC and JPL Archives.

5. Hyron Spinrad, informal conversation with Schorn, n.d.

6. Key Voyager/Saturn encounter literature includes Edward C. Stone, "The Voyager Encounters with Saturn," *AIAA Journal* 22 (1984): 498; T. Gehrels and M. Matthews, eds., *Saturn* (Tucson: University of Arizona Press, 1984); Edward C. Stone, "The Voyager Mission: Encounters with Saturn," *Journal of Geophysical Research* 88 (1983): 8369; and Edward C. Stone and Ellis D. Miner, "Voyager 2 Encounter with the Saturnian System," *Science* 215 (1982): 499. Many of these publications give references to what was known or suspected about the Saturnian system prior to the Voyager encounters.

7. The existence of Saturn's fierce equatorial winds was known as early as 1903. See, for example, Alexander, *Planet Saturn,* 279–83. Interestingly enough, most of the relevant observations were made by *amateur* astronomers with relatively small telescopes.

8. After the Moon and Mars, Saturn is arguably the most photogenic and popular object in the universe with the general public, and professional planetary astronomers are not immune to that lure; consequently the literature on the ringed planet is enormous. Selected references include the standard on Saturn research through the 1950s, as mentioned in note 1, Alexander, *Planet Saturn.*

A good, concise, summary of Saturn research from the earliest times is in David Morrison, *Voyages to Saturn,* NASA SP-451, Washington, D.C., 1982. Morrison's main theme is the Voyager encounters, but his summary of findings through the Pioneer 11 flyby of Saturn tells a complicated story in a short, understandable, yet readable form. The unremitting question—"How many satellites does Saturn really have?"—was addressed by Stephen Larson and John W. Fountain, *Sky & Telescope,* November 1980, 356, and is only one of many articles that address the question, but it is representative.

9. The observations needed to investigate this problem were difficult, and there weren't that many astronomers willing to devote their personal time and telescope time to the effort. It is most illustrative to see how JPL approached this problem, and some of the details can be found in Box 1, Folder 5, HC 105, JPL Archives. A memorandum from E. Davis and T. Gindorf to R. Miles, November 7, 1972, discusses planned observations—basically, observing runs at Mount Wilson and Palomar Observatories. There follow detailed plans, with dates precisely laid out, flow diagrams, follow up memos, cost estimates, a report on a workshop on the topic, progress reports, and suggestions for future observational and theoretical work.

To an observational astronomer, the amount of money and effort JPL devoted to this problem seems enormous compared to the relatively minor series of observations—just the preparation and distribution of the various memos involved must have involved a lot of time and money. In addition, mission planners of course needed definite information by a definite date, but in this case the tendency of managers to assume that almost any problem can be resolved in a predictable and timely manner ran up against the fact that the universe is not that cooperative. Astronomers had learned this truism a long time ago, and they learned it the hard way.

10. Morrison, *Voyages to Saturn,* gives a concise summary of Pioneer results at Jupiter and Saturn on pages 14–28. Standard collections of technical papers are the special January 25, 1980, issue of *Science* 207:400–453, and the special November 1, 1980, issue of the *Journal of Geophysical Research* 85:5651–5958. As has been the case with every successful spacecraft encounter with another planet, research papers in the form of preliminary accounts, summaries as of specific times, follow-up studies, and retrospectives are widely scattered through a bewildering variety of publications, and just their citations would form a respectable volume.

11. A lot of test material exists in the JPL Archives. Just one example is Document 618-515, Voyager Mission Operations Test Plan, Volume III-C, Flight Team Test and Training, Saturn Encounter, April 1980, JPL 93, Box 2, Folder 28, spells out in detail goals and responsibility.

12. Miner interview. Miner later became associated with the Cassini mission to Saturn and, at the time of the interview, was science advisor for that project.

13. We eschew any attempt at a complete listing of popular or technical reports of the Voyagers' encounters with Saturn. Morrison, *Voyages to Saturn,* gives a blow-by-blow account of both encounters from the viewpoint of a planetary scientist who was there at the time and honestly "tells it like it was." Notable collections of technical papers regarding the Voyager 1 encounter appear in the special April 10, 1981, issue of *Science* 212:159–243, and the special August 20, 1981, issue of *Nature* 292:675–755. The Voyager 2 encounter is in the special January 29, 1982, issue of *Science* 215:499–594, while results from both flybys are in *Journal of Geophysical Research* 88:8625–9018, 1983. There are an enormous number of follow-up studies based on Voyager results, but here, as in the cases of all the planetary encounters, we limit ourselves to the what was learned during and in the immediate aftermath of the flybys, for developments in the subsequent decades deserve another book. We must, however, mention Henry F. Cooper Jr.'s article that begins on page 39 of the August 24, 1981, issue of the *New Yorker;* he was there and he really captured "the spirit of the times." This is the most accurate and readable account of a spacecraft's encounter with a planet that we have had the pleasure to read.

14. The application of spiral density waves to Saturn's rings was first suggested in P. Goldreich and S. Tremaine, *Icarus* 34 (1978): 240. The original idea goes back to Bertil Lindblad decades before; see, for example, his article (along with Per Olaf Lindblad) "On Rotating Ring Orbits in Galaxies," in *International Astronomical Union Symposium No. 5, Comparison of the Large-Scale Structure of the Galactic System with That of Other Stellar Systems,* ed. N. G. Roman (Cambridge: Cambridge University Press, 1958), 8. See also "Fact Sheet: Voyager Saturn Science Summary," at www.jpl.nasa.gov, March 13, 1998.

15. Ellis Miner to Henry C. Dethloff, August 28, 2000.

16. Ibid.

17. And see "Fact Sheet: Voyager Saturn Science Summary."

18. Ibid.

19. Alexander, *Planet Saturn,* has illustrations of such dark spokes on page 256 and in plate I, though in these cases the features appear on the A ring! Perhaps spokes appear sometimes on one ring and sometimes on another; as the cause and nature of these features is a mystery, it would seem unwise to rule out that possibility. There were many radial spokes that maintained their radical character through most of

an orbit, leading to the conclusion that their tiny particles were charged sufficiently to cause them to be frozen in the magnetic field, at least temporarily.

20. For a brief summation of the Titan encounter, see "Fact Sheet: Voyager Saturn Science Summary."
21. Donald M. Hunten, NASA Conference Publication 2068, 1978, 127. We asked Hunten what it felt like to have his predictions come true. "It was great!" he replied, for the life of a theoretician is not easy.
22. Stone interview, January 28, 1999.
23. Edward C. Stone, *Nature*, August 20, 1981, 678. Somewhat confusingly, this reference appears in a special issue of *Nature* dedicated to the Voyager 1 encounter with Saturn.
24. Miner to Dethloff, August 28, 2000.
25. Raymond N. Batson, *Voyager 1 and 2 Atlas of Six Saturnian Satellites* (Washington, D.C.: NASA, 1984), has almost no interpretation, but presents readable maps and large selections of images for Mimas, Enceladus, Tethys, Dione, Rhea, and Iapetus. Titan's surface was hidden from the Voyagers' view, and the observations of Phoebe and Hyperion were so poor that no meaningful maps could be drawn.
26. Miner to Dethloff, August 28, 2000. We interviewed a number of cognizant JPL project engineers on this subject, and no one could come up with a definitive answer. One finally attributed it to the "Great Galactic Ghoul."

11. URANUS, NEPTUNE, AND BEYOND

1. Uranus is often called "blue," "bluish," or "blue-green" in the literature, but in our experience, most observers who have actually looked visually at the planet through telescopes—be the instruments small or large—call it "green." Perceived color is, to some extent, subjective, and often affected by preconceptions. Mars, for example, is called the "Red Planet," even though it is actually dull, drab, and yellowish, much like the true color of Arizona's Painted Desert, which reveals itself at high noon. Moreover, there are celestial objects known as "planetary nebulae," because some of them resemble Uranus by appearing round and greenish.
2. What might be called the "classical" knowledge of Uranus is well and fully described in A. F. O'D. Alexander, *The Planet Uranus: A History of Observation, Theory, and Discovery* (New York: American Elsevier Publishing, 1965). What we knew of the Uranian system just a few years before the Voyager 2 encounter is concisely summarized in Garry Hunt, ed., *Uranus and the Outer Planets* (Cambridge: Cambridge University Press, 1982). This volume also contains several interesting articles concerning the planet's discovery. Jay T. Bergstrahl, ed., *Uranus and Neptune* (Washington, D.C.: NASA, 1984), contains a wide ranging survey of our knowledge of those two planets shortly before Voyager 2 visited them.
3. Ibid.
4. Ibid.
5. The discovery of rings around Uranus was a big news story and, for example, hit the front page of the *New York Times* on March 31, 1977.
6. The most complete discovery paper is J. L. Elliot, E. W. Dunham, and D. J. Mink in *Nature* (1977): 267, 328. Elliot reviews stellar occultation results as they bear on the Solar System in Geoffrey Burbidge, David Layzer, and John G. Phillips, eds., *Annual Review of Astronomy and Astrophysics* (Palo Alto, Calif.: Annual Reviews,

1979), vol. 17. Hunt, *Uranus,* contains articles on the rings or Uranus by Elliot as well as A. Brahic.

7. Miner to Dethloff, August 28, 2000.

8. The two most important sources are the special July 4, 1986, issue of *Nature,* and *Journal of Geophysical Research* 92 (1987): 14,873–15,375. As is the case with other Voyager encounters, there are a large number of other special issues on the subject in various journals, and there are, of course, literally hundreds (perhaps thousands, for they are still appearing) of individual articles in scientific, engineering, and popular publications. Then too, there are various symposium, "workshop," and meeting volumes, and one can go to many a bookstore and find a number of volumes concerning the Voyagers, the Solar System, space exploration, and the like, in which the Uranus results are presented.

9. Ibid.

10. *Journal of Geophysical Research* 92 (1987): 14,873–15,375; and see *Nature,* July 4, 1986. The Uranus encounter was overshadowed in the national media by the explosion and tragedies of the space shuttle Challenger on January 26.

11. Ibid.

12. The optical occultations were observations of a star as the Uranus rings passed in front of the star. Miner to Dethloff, August 20, 2000.

13. Ibid.; "Fact Sheet: Uranus Science Summary," http://www.jpl.nasa.gov, March 17, 1998, pp. 1–6.

14. Ibid.

15. Ibid.

16. Ibid.

17. Ibid.

18. Ibid.

19. Ibid.

20. Ibid.

21. Ridenoure interview.

22. Ibid.

23. Kolhase, *Voyager Neptune Travel Guide,* gives a superb summary of our knowledge of Neptune, the history of the Voyager program, the status of Voyager 2, and plans for the Neptune flyby, all at the time when encounter observations were beginning. This concise, well-written handbook was produced as a handout for reporters covering the mission and is a model of its kind.

24. See, for example, H. J. Reitsema et al., *Science* 215 (1982): 289; C. E. Couvault et al., *Icarus* 67 (1986): 126; and W. B. Hubbard et al., *Nature* 319 (1986): 639.

25. Two standard references are the special December 15, 1989, issue of *Science,* and the *Journal of Geophysical Research* 96 (1991): 18,903–19,268.

26. See Kolhase, *Voyager Neptune Travel Guide;* and see Edward C. Stone, "The Neptune Challenge: Many Answers and More Questions," *Geophysical Research Letters* 15, no. 10 (1990): 1643; Edward C. Stone, "Voyager at Neptune," *Engineering and Science* 53, no. 3 (1990): 24.

27. Ibid.; and see Edward C. Stone and Ellis D. Miner, "Results from the Voyager 2 Flyby of Neptune in August of 1989," *McGraw-Hill Yearbook of Science and Technology, 1989* (New York: McGraw-Hill, 1989), and Edward C. Stone and Ellis D. Miner, "The Voyager 2 Encounter with the Neptunian System," *Science* 246 (1989): 1417.

28. Ibid.

12. TERMINATION SHOCK

1. "Voyager's Interstellar Mission Description," web site information download, www.jpl.nasa.gov, March 8, 1998, NASA HRC.
2. Nick Flowers, "Striking the Solar Shock Wave," *New Scientist,* March 22, 1997, 41.
3. Stone interview, January 28, 1999.
4. Ibid.
5. Stone interview, January 28, 1999; Ridenoure interview; and see www.photojournal. jpl.nasa.gov.Solar System/Voyager/Montage.
6. As the Voyager Interstellar Mission is still in progress, it is impossible to give any comprehensive references to the results of that project. There have been a huge number of news items, scientific papers, summaries to date, and the like since the Neptune encounter, and we mention only a few here. Most useful was an interview with Edward C. Stone Jr., the current Voyager project scientist (and current director of JPL), by Dethloff and Schorn on January 28, 1999. A fine interim progress report by Nick Flowers appeared in *New Scientist* for March 22, 1997. Readers on the world wide web can find the most recent information: www.jpl.nasa.com. See, for example, "*Voyager's* Interstellar Mission," March 8, 1998, and "Voyager Approach to Maintaining Science Data Acquisition for a 30 Year Extended Mission," May 19, 2000. The latter reference gives a detailed picture of the sequential shutdowns of various Voyager systems in an attempt to prolong the lifetimes of these probes.
7. Massey interview.
8. Ibid.
9. Ibid.
10. Ibid.
11. JPL Media Relations Office, "Dynamic Terrain and Volcanoes Galore on Io," May 31, 2000, on line at www.jpl.nasa.gov. Transmission of data from the Galileo orbiter, especially the transmission of video images, is dramatically slower than planned, due to the failure of a large, "high-gain," directional antenna to deploy fully. However, results are coming in, and to note just one recent example, the *Houston Chronicle* for May 19, 2000, reported on the incredibly high level of activity on Io and the huge (by terrestrial standards) tides that deform the satellite's solid body by hundreds of feet every forty-two hours.
12. Publicity concerning Cassini is understandably at a low level during the cruise phase of the mission. a "Cassini Fact Sheet" issued by JPL just before the probe's launch provides a succinct summary of the planned mission.
13. "Gee Whiz Stuff About the Voyager Project," web site information download, www.jpl.nasa.gov, March 8, 1998, NASA HRC.
14. Stephen P. Maran, interview with Schorn, January 7, 1999, NASA HRC; David Morrison, interview with Schorn, Washington, D.C., January 7, 1999, NASA HRC.
15. Morrison interview.
16. Ibid.
17. Frank J. Collela, JPL Public Information Officer, to Bruce C. Murray, JPL Director, December 17, 1980, JPL 89-13, folder 100, JPL Archives. In the same folder is a 1977 letter from R. A. Mills, the Voyager Project Manager at NASA Headquarters to John Casani, the project manager for the Voyager Project at JPL, lamenting the fact that there are no funds available to support public affairs activities in connection with Voyager, but at the same time acknowledging the intense public interest in the subject.

18. *Star Trek: The Motion Picture* is copyrighted 1980 by Paramount Pictures. At this writing, the film is available on video in a "six-pack" of films that feature the original cast from the first television series.
19. Kelly Carey, Saturn Customer Assistance Center, to Susan Schorn, e-mail, May 20, 2000. Courtesy Susan Schorn.
20. Smith interview.
21. Ibid.
22. David Crawford, interview with Schorn, Austin, Tex., January 7, 1999, NASA HRC.
23. Ibid.
24. David Chandler, interview with Schorn, Austin, Tex., January 9, 1999, NASA HRC.
25. Stone interview, January 28, 1999.
26. Miner interview.
27. "Voyager Mission Operations Status Report," no. 1119, April 8–April 14, 2000, on line at www.jpl.nasa.gov; and see "Voyager 1 Now Most Distant Human-Made Object in Space."

BIBLIOGRAPHY

Particularly valuable resources for the history of the Voyager Grand Tour include the Jet Propulsion Laboratory Archival and Records Center Collections located in Pasadena, California; the NASA Historical Records Collections located in the NASA History Office, NASA Headquarters, Washington, D.C.; and collected oral history interviews that have been placed both in the NASA Historical Records Collections and the Jet Propulsion Laboratory Archives as a part of this history project. Other relevant oral history interviews previously collected are located in the JPL Archives, the NASA Historical Records Collections, and the Planetary Astronomy Papers at Texas A&M University Archives.

Jet Propulsion Laboratory archival materials are readily accessed through registers and indexes. Voyager-related JPL archival materials are found in the Historical Collections, the Space Flight Collections (Voyager Project Office Collections), and the Charles E. Kohlhase Collection. The "Mariner Jupiter/Saturn 1977" and "Grand Tour" documents are key components of the Voyager Collections, as are the Sun Sensor, Guidance and Control, Telemetry Subsystem, Space Flight Operations, and science and encounter subsystem collections.

An interesting component of the JPL archival collections are the lecture series, training, orientation, and information videos. These documents provide a vivid picture of what it was really like to design, build, and operate the Voyagers. However, their enormous amount of detail, combined with a confusing multitude of acronyms, made it impossible to use them more fully without resorting to extremely long and involved explanations.

A bibliography of key books and documents follows, along with a listing of journals, proceedings, and reports. Useful bibliographic compilations include *Voyager: A Bibliography*, published in 1980 by Maryruth Glogowski and Pamela G. Kobelski of the State University of New York at Buffalo for the Vance Bibliographies Series. An important electronic *Voyager Bibliography of Scientific Publication* can be accessed through JPL. For information contact Richard Rudd at richard.rudd@jpl.nasa.gov. Jet Propulsion Laboratory Voyager news releases can also be accessed on line at http:www.jpl.nas.gov. In addition, access to numerous relevant records and materials is available on line through the History Division home page at http://history.nasa.gov.

Among the most essential published books and documents are Clayton R. Koppes, *JPL and the American Space Program: A History of the Jet Propulsion Laboratory* (1982); Ronald A. Schorn, *Planetary Astronomy: From Ancient Times to the Third Millennium* (1998); and David W. Swift, *Voyager Tales: Personal Views of the Grand Tour* (1998). Douglas J. Mudgway, *Uplink-Downlink: A History of the Deep Space Network* (2001) provides numerous studies of the critical interface between the DSN and the Voyager missions to the outer planets. Key published documents include the reports and symposia of the National Academy of Sciences–Space Science Board, and the electronically distributed JPL Voyager Mission Reports, Goddard Space Flight Center science reports (nand@voyager.gsfc.nasa.gov), and NASA news releases. A particularly important and useful collection of documents is *Exploring the Unknown: Selected Documents in the History of the U.S. Civilian Space Program,* volumes 1–3, published in the NASA History Series between 1995 and 1998 (NASA SP-4407). The proceedings of the various conferences held by the Lunar and Planetary Science Institute in Houston, Texas (issued by a variety of publishers), contain many papers relating to Voyager investigations. Among the leading journals providing technical articles relating to Voyager missions are *Science, Nature, Icarus,* and the *Journal of Geophysical Research.* Many popular articles can be found in such magazines as *Aviation Week & Space Technology, Astronomy,* and *Sky & Telescope.*

The special issues of scientific journals that are devoted to individual Voyager planetary encounters are most useful references on the subjects. They are listed in a separate section at the end of this bibliography.

The documents mentioned here are only a selection of what has been published. In particular, there have been literally hundreds of books written in the English language alone dealing with what the Voyagers revealed about the outer Solar System, and thousands of scientific and technical articles in various publications. Here we only give a sampling that we believe is representative and most relevant. Our aim in this bibliography is neither to provide a comprehensive manual on how to design, build, and operate planetary probes nor to write an encyclopedia of our knowledge of the outer Solar System. Instead, we hope to provide readers with useful references as to how the Voyagers came to be, what they did, and what they promise for the future.

Critical elements of our research have been the oral history interviews that are available variously in the JPL Archives, the NASA Historical Records Collections, and the Planetary Astronomy Papers located in the Texas A&M University Archives. Interviews are listed alphabetically by name of interviewee.

INTERVIEWS

Unless otherwise stated, interviews were conducted by the authors.

Bergstrahl, Jay T. Washington, D.C., February 23, 1993.
Casani, John Richard. Pasadena, Calif., January 28, 1999.
Chamberlin, Joseph W. Houston, Tex., April 13, 1993.
Chandler, David. Austin, Tex., January 9, 1999.

Cochran, Bill. Austin, Tex., June 1, 1993.
Cochran, Anita. Austin, Tex., June 1, 1993.
Cortright, Edgar. With R. Cargill Hall. Washington, D.C., March 4, 1968. JPL Archives.
Crawford, David. Austin, Tex., January 7, 1999.
Cunningham, Newton. With R. Cargill Hall. Washington, D.C., March 6, 1968. JPL Archives.
de Vaucouleurs, Gerard. Austin, Tex., June 2, 1993.
Draper, Ronald F. Pasadena, Calif., January 29, 1999.
Franklin, Fred A. Boston, Mass., April 22, 1993.
Heacock, Raymond Leroy. Pasadena, Calif., February 4, 1999.
Jeffries, William. Austin, Tex., June 1, 1993.
Kohlhase, Charles E. Pasadena, Calif., January 25, 1999.
Lane, Arthur Lonne. Pasadena, Calif., October 1, 1999.
Malina, Frank J. With R. Cargill Hall. Pasadena, Calif., October 29, 1968. JPL Archives.
Maran, Stephen P. Austin, Tex., January 7, 1999.
Marsden, Brian. Boston, Massachusetts, April 21, 1993.
Miner, Ellis D. Pasadena, Calif., January 26, 1999.
Morrison, David. Austin, Tex., January 7, 1999.
Nicks, Oran Wesley. With R. Cargill Hall. Washington, D.C., August 26, 1968. JPL Archives.
Ridenoure, Rex Wade. Pasadena, Calif., January 27, 1999.
Schneiderman, Dan. Pasadena, Calif., July 17, 1982. JPL Archives.
Schurmeier, Harris (Bud). Pasadena, Calif., September 25, 1970.
Sinton, William M. Phoenix, Ariz., January 4, 1993.
Smith, Bradford A. Austin, Tex., January 8, 1999.
Stone, Edward C. Pasadena, Calif., January 28, 1999.
Textor, George P. Pasadena, Calif., February 2, 1999.
Trafton, Larry. Austin, Tex., June 2, 1993.
Traub, Wesley. Boston, Mass., April 22, 1993.
Tull, Robert. Austin, Tex., June 1, 1993.
Whipple, Fred L. Boston, Mass., April 20, 23, 1993.

ARCHIVAL AND MANUSCRIPT COLLECTIONS

Kuiper, Gerard P. Manuscripts. Special Collections, University of Arizona, Tucson.
NASA, Jet Propulsion Laboratory Archives and Records. Pasadena, California.
NASA, Jet Propulsion Laboratory Papers. NASA Historical Reference Collections, NASA History Office, National Aeronautics and Space Administration, Washington, D.C.
NASA, Lunar and Planetary Astronomy Papers. NASA Historical Reference Collections, NASA History Office, National Aeronautics and Space Administration, Washington, D.C.
National Science Foundation. Miscellaneous Papers. National Science Foundation, Washington, D.C.
Planetary Astronomy Papers. Texas A&M University Archives, College Station.

BOOKS AND DOCUMENTS

Akens, David S. *Historical Origins of the George C. Marshall Space Flight Center.* MSFC Historical Monograph No. 1. Huntsville, Ala.: Marshal Space Flight Center, December 1960.

Alexander, A. F. O'D. *The Planet Saturn: A History of Observation, Theory, and Discovery.* London: Faber and Faber, 1962.

———. *The Planet Uranus: A History of Observation, Theory, and Discovery.* New York: American Elsevier, 1965.

Ambrose, Stephen E. *Rise to Globalism: American Foreign Policy since 1938.* 5th ed. New York: Viking Penguin Books, 1988.

Beatty, J. Kelly, Brian O'Leary, and Andrew Chaikin, eds. *The New Solar System.* 2d ed. Cambridge, Mass.: Sky Publishing; Cambridge: Cambridge University Press, 1982. As of 2000, this work is in its fourth edition.

Bergstrahl, Jay T., ed. *Uranus and Neptune.* NASA CP-2330. Pasadena, Calif., 1984.

Bergstrahl, Jay T, Ellis D. Miner, and Mildred Mathews Shapley, eds. *Uranus.* Tucson: University of Arizona Press, 1991.

Blanco, V. M., and S. W. McCuskey. *Basic Physics of the Solar System.* Reading, Mass: Addison-Wesley, 1961.

Bobrovnikoff, Nicholas T. Edited by Roger B. Culver and David D. Meisel. *Astronomy before the Telescope.* Vol. 1, *The Earth-Moon Systems.* Tucson: Pachart Publishing, 1984.

———. *Astronomy before the Telescope.* Vol. 2, *The Solar System.* Tucson: Pachart Publishing, 1984.

Burgess, Eric. *By Jupiter: Odysseys to a Giant.* New York: Columbia University Press, 1982.

———. *Uranus and Neptune: The Distant Giants.* New York: Columbia University Press, 1988.

Butrica, Andrew J. *To See the Unseen: A History of Planetary Radar Astronomy.* NASA SP-4218. Washington, D.C., 1996.

Cooper, Henry S. F., Jr. *Imaging Saturn: The Voyager Flights to Saturn.* New York: Holt, Rinehart and Winston, 1982.

Corliss, William R. *A History of the Deep Space Network.* NASA CR-151915. May 1, 1976.

Cruikshank, Dale P., ed. *Neptune and Triton.* Tucson: University of Arizona Press, 1995.

Dethloff, Henry C. *Suddenly . . . Tomorrow Came: A History of the Johnson Space Center.* NASA SP-4307. Washington, D.C., 1993.

———. *The U.S. and the Global Economy Since 1945.* New York: Harcourt, Brace, 1997.

de Vaucouleurs, Gerard. *Geometric and Photometric Parameters of the Terrestrial Planets.* Memorandum RM-4000. Santa Monica, Calif.: Rand Corporation, March 1964.

———. *Physics of the Planet Mars.* London: Faber and Faber, 1954.

———. *The Planet Mars.* 2d ed. London: Faber and Faber, 1952.

Emme, Eugene M. *Aeronautics and Astronautics: An American Chronology of Science and Technology in the Exploration of Space, 1915–1960.* Washington, D.C.: NASA, 1961.

Ezell, Edward Clinton, and Linda Neuman Ezell. *On Mars: Exploration of the Red Planet, 1958–1978.* NASA SP-4212. Washington, D.C., 1984.

Ezell, Linda Neuman. *NASA Historical Data Book.* Vols. 1–3. NASA SP-4012. Washington, D.C., 1988.

Gehrels, Tom, and Mildred Shapley Mathews, eds. *Saturn.* Tucson: University of Arizona Press, 1984.

Glennan, T. Keith. *The Birth of NASA: The Diary of T. Keith Glennan.* NASA SP-4501. Washington, D.C., 1993.

Gordon, Theodore J., and Julian Scheer. *First into Outer Space.* New York: St. Martin's Press, 1959.

Greely, Ronald, and Raymond Batson. *The NASA Atlas of the Solar System.* New York: Cambridge University Press, 1996.

Greenberg, Richard, and Andre Brahic, eds. *Planetary Rings.* Tucson: University of Arizona Press, 1984.

Grosser, Martin. *The Discovery of Neptune.* Cambridge: Harvard University Press, 1962.

Hall, R. Cargill. *Lunar Impact: A History of Project Ranger.* Washington, D.C.: GPO, 1980.

Hansen, James R. *Engineer in Charge: A History of the Langley Aeronautical Laboratory 1917–1958.* NASA SP-4305. Washington, D.C.: NASA Scientific and Technical Office, 1987.

Hinks, Arthur R. *Astronomy.* New York: Henry Holt Williams and Northgate, April 1911.

Hunt, Garry, and Patrick Moore. *Atlas of Neptune.* Cambridge: Cambridge University Press, 1994.

———. *Atlas of Uranus.* Cambridge: Cambridge University Press, 1989.

———. *Jupiter.* New York: Rand McNally, 1981.

———. *Saturn.* New York: Rand McNally, 1982.

Johnson, Lyndon B. *The Vantage Point: Perspectives of the Presidency, 1963–1969.* New York: Rinehart & Winston, 1971.

Kaufmann, William J., III. *Exploration of the Solar System.* New York: Macmillan; London: Collier Macmillan, 1978.

Kluger, Jeffrey. *Journey Beyond Selene: Remarkable Expeditions Past Our Moon and to the Ends of the Solar System.* New York: Simon & Schuster, 1999.

Koenig, L. R., et al. *Handbook of the Physical Properties of the Planet Venus.* NASA SP-3029. Washington, D.C., 1967.

Kohlhase, Charles. *The Voyager Neptune Travel Guide.* JPL Publication 89-24. Pasadena, Calif., June 1, 1989.

Koppes, Clayton R. *JPL, and the American Space Program: A History of the Jet Propulsion Laboratory.* New Haven, Conn.: Yale University Press, 1982.

Kraemer, Robert S. *Beyond the Moon: A Golden Age of Planetary Exploration, 1971–1978.* Washington, D.C.: Smithsonian Institution Press, 2000.

Kuiper, Gerard P., ed. *The Atmospheres of the Earth and Planets.* Rev. ed. Chicago: University of Chicago Press, 1952.

———. *The Earth as a Planet.* Vol. 2 of *The Solar System.* 2d impression. Chicago: University of Chicago Press, 1958.

———. *The Sun.* Vol. 1 of *The Solar System.* 2d impression. Chicago: University of Chicago Press, 1954.

Kuiper, Gerard P., and Barbara M. Middlehurst, eds. *Planets and Satellites.* Vol. 3 of *The Solar System.* Chicago: University of Chicago Press, 1961.

Launius, Roger D., John M. Logsdon, and Robert W. Smith, eds. *Reconsidering Sputnik: Forty Years since the Soviet Satellite.* New York: Harwood Academic Press, 2000.

Leslie, Stuart W. *The Cold War and American Science: The Military-Industrial-Academic Complex at MIT and Stanford.* New York: Columbia University Press, 1993.

Levine, Arnold S. *Managing NASA in the Apollo Era.* NASA SP-4102. Washington, D.C., 1982.

Levitt, I. M. *A Space Traveler's Guide to Mars.* New York: Henry Holt, 1956.

Littman, Mark. *Planets Beyond: Discovering the Outer Solar System.* New York: John Wiley & Sons, 1988.

Logsdon, John M., gen. ed. *Exploring the Unknown: Selected Documents in the History of the U.S. Civilian Space Program.* Vol. 1, *Organizing for Exploration;* vol. 2, *External Relationships;* vol. 3, *Using Space.* Washington, D.C.: NASA History Series, 1995–98.

Mack, Pamela, ed. *From Engineering Science to Big Science: The NACA and NASA Collier Trophy Research Project Winners.* NASA SP-4219. Washington, D.C., 1989.

Mason, Brian. *Meteorites.* New York: John Wiley and Sons, 1962.

McCurdy, Howard E. *Inside NASA: High Technology and Organizational Change in the U.S. Space Program.* Baltimore: Johns Hopkins University Press, 1987.

McDougall, Walter A. . . . *the Heavens and the Earth: A Political History of the Space Age.* New York: Basic Books, 1985.

McKinley, D. W. R. *Meteor Science and Engineering.* New York: McGraw-Hill, 1961.

Middlehurst, Barbara M., and Gerard P. Kuiper, eds. *The Moon, Meteorites, and Comets.* Vol. 4 of *The Solar System.* Chicago: University of Chicago Press, 1963.

Miner, Ellis D. *Uranus: The Planet, Rings and Satellites.* 2d ed. New York: John Wiley & Sons, 1998.

Mizwa, Stephen P., ed. *Nicholas Copernicus: A Tribute of Nations.* New York: Kosciuszko Foundation, 1945.

Moore, Patrick. *The Planet Neptune.* Chichester, England: Ellis Horwood, 1988.

———. *The Planet Venus.* 2d ed. New York: Macmillan, 1959.

Morrison, David. *Voyages to Saturn.* NASA SP-451. Washington, D.C., 1982.

Morrison, David, and Jane Samz. *Voyage to Jupiter.* NASA SP-439. Washington, D.C., 1980.

Moulton, Forest Ray. *An Introduction to Celestial Mechanics.* 2d rev. ed. New York: Macmillan, 1959.

Mudgway, Douglas J. *Uplink-Downlink: A History of the Deep Space Network, 1957–1997.* NASA SP-2001-4227. Washington, D.C., 2001.

Murray, Bruce. *Journey into Space: The First Three Decades of Space Exploration.* New York: W. W. Norton, 1989.

National Air and Space Museum. Space Science and Exploration Department. *Space History Oral History Project Catalogue.* Washington, D.C.: Smithsonian Institution, 1985.

Naugle, John E. *First among Equals: The Selection of NASA Space Science Experiments.* NASA SP-4215. Washington, D.C., 1991.

Newburn, R. L., Jr., and S. Gulkis. *A Brief Survey of the Outer Planets Jupiter, Saturn, Uranus, Neptune, Pluto, and Their Satellites.* Technical Report 32-1529. Pasadena, Calif.: Jet Propulsion Laboratory, April 15, 1971.

Newell, Homer E. *Beyond the Atmosphere: Early Years of Space Science.* NASA SP-4211. Washington, D.C., 1980.

Nicks, Oran W. *Far Travelers: The Exploring Machines.* NASA SP-480. Washington, D.C., 1985.

Nimmen, Jane Van, Leonard C. Bruno, and Robert L. Rosholt. *NASA Historical Data Book.* Vol. 1, *NASA Resources 1958–1968.* NASA SP-4012. Washington, D.C., 1988.

Ninninger, H. H. *Out of the Sky: An Introduction to Meteoritics.* Unaltered reproduction of first edition. New York: Dover Publications, 1959.

Osterbrock, Donald E. *James E. Keeler: Pioneer American Astrophysicist and the Early Development of American Astrophysics.* Cambridge: Cambridge University Press, 1984.

Osterbrock, Donald E., John R. Gustafson, and W. J. Shiloh Unruh. *Eye on the Sky: Lick Observatory's First Century.* Berkeley and Los Angeles: University of California Press, 1988.

Pash, Boris T. *The Alsos Mission.* First Award printing. New York: Award Books; London: Tandem Books, 1970.

Payne-Gaposchkin, Cecilia. *Introduction to Astronomy.* New York: Prentice-Hall, 1954.

Peek, Bertrand M. *The Planet Jupiter.* London: Faber and Faber, 1958.

Proctor, Richard A. *Other Worlds than Ours: The Plurality of Worlds Studied under the Light of Recent Scientific Researches.* Reprinted from the latest London edition. New York: Hurst & Company, n.d. Published after 1873, the year of the author's death. There is an author's preface from the fourth edition.

Richter, Nikolaus B. *The Nature of Comets.* Translated and rev. ed. by Arthur Beer. London: Methuen, 1963.

Ride, Sally, and Tam O'Shaughnessy. *Voyager: An Adventure to the Edge of the Solar System.* New York: Crown Publications, 1999.

Rogers, John H. *The Giant Planet Jupiter.* Cambridge: Cambridge University Press, 1995.

Roland, Alex. *Model Research: The National Advisory Committee for Aeronautics, 1915–1958.* NASA SP-4103. Washington, D.C., 1985.

Rosholt, Robert L. *An Administrative History of NASA, 1958–1963.* Washington, D.C.: GPO, 1966.

Roth, Gunter D. *The System of Minor Planets.* Princeton, N.J.: Nostrand, 1962.

Russell, Henry Norris, Raymond Smith Dugan, and John Quincy Stewart. *Astronomy: A Revision of Young's Manual of Astronomy.* Vol. 1, *The Solar System.* Rev. ed. Boston: Ginn, 1945.

Sagan, Carl, Tobias C. Owen, and Smith, Harlan J., eds. *Planetary Atmospheres: International Astronomical Union Symposium No. 40, Held in Marfa, Texas, U.S.A., October 26–31, 1969.* Dordrecht, Holland: D. Reidel; New York: Springer Verlag, 1971.

Schorn, Ronald A. *Planetary Astronomy: From Ancient Times to the Third Millennium.* College Station: Texas A&M University Press, 1998.

Slipher, Earl C. *The Photographic Story of Mars.* Cambridge, Mass.: Sky Publishing; Flagstaff, Ariz.: Northland Press, 1962.

Space Science Board, Task Group on Planetary and Lunar Exploration, *Space Science in the Twenty-First Century: Imperatives for the Decades 1995–2015.* Washington, D.C.: National Academy Press, 1988.

Swarup, G., A. K. Bag, and K. S. Shukla, eds. *History of Oriental Astronomy.* Proceedings of International Astronomical Union Colloquium No. 91. New Delhi, India, November 13–16, 1985. Cambridge: Cambridge University Press, 1987.

Swift, David W. *Voyager Tales: Personal Views of the Grand Tour.* Reston, Va.: American Institute of Aeronautics and Astronautics, 1997.

Swings, Pol. "Venus Through a Spectroscope." Mimeographed review monograph. November 1968. Author's papers.

Tatarewicz, Joseph N. *Space Technology & Planetary Astronomy.* Bloomington: Indiana University Press, 1990.

Urey, Harold C. *The Planets: Their Origin and Development.* New Haven, Conn.: Yale University Press; London: Oxford University Press, 1952.

Washburn, Mark. *Distant Encounters: The Exploration of Jupiter and Saturn.* San Diego: Harcourt Brace Jovanovich, 1983.

Watson, Fletcher G. *Between the Planets.* 1st ed. Philadelphia: Blakiston, 1945.
———. *Between the Planets.* Rev. ed. Garden City, N.Y.: Doubleday, 1962.
Webb, W. L. *Brief Biography and Popular Account of the Unparalleled Discoveries of T. J. J. Lee.* Lynn, Mass.: Thomas P. Nichols, 1913.
Whipple, Fred L. *Earth, Moon, and Planets.* 1st ed. Philadelphia: Blakiston, 1947.
Whitaker, Ewen A. *The University of Arizona's Lunar and Planetary Laboratory: Its Founding and Early Years.* Tucson: University of Arizona Printing-Reproductions Department, [c. 1975].
Whitehead, A. Bruce. *The Photomosaic Globe of Mars.* Pasadena, Calif.: Jet Propulsion Laboratory, n.d.
Wolf, Max. *Max Wolf, 1863–1932: Ein Gedenkblatt.* Berlin: Walter de Gruyter, 1933.
Wood, John A. *Meteorites and the Origin of Planets.* New York: McGraw-Hill, 1968.
Wright, Helen. *Sweeper in the Sky: The Life of Maria Mitchell, First Woman Astronomer in America.* New York: Macmillan, 1949.
Yeomans, Donald K. *Comets: A Chronological History of Observation, Science, and Folklore.* New York: John Wiley & Sons, 1991.

ARTICLES, PROCEEDINGS, REPORTS, AND SYMPOSIUMS

Aeronomy of CO₂ Atmospheres: Proceedings of the Fifth Arizona Conference on Planetary Atmospheres. Reprinted from *Journal of Planetary Atmospheres* 28 (September 1971): 833–1086.
Alfven, Hannes, and Gustaf Arrhenius. "Structure and Evolutionary History of the Solar System, Part III." Preprint of an article in press for *Astrophysics and Space Science* 12 (1973). Author's papers.
Anderson, Linda. "Mars-Surface Features and Atmosphere." 2 vols. Jet Propulsion Laboratory Literature Search No. 667. August 2, 1965. Author's papers.
———. "Venus and Other Planets." Jet Propulsion Laboratory Literature Search No. 688. December 30, 1965. Author's papers.
The Atmosphere of Venus: Papers from the Second Arizona Conference on Planetary Atmospheres. Reprinted from *Journal of the Atmospheric Sciences* 25 (July 1968): 553–671.
Averner, M. M., and R. D. MacElroy. *On the Habitability of Mars: An Approach to Planetary Ecosynthesis.* Ames Research Center. NASA SP-414. Washington, D.C., 1976.
Butrica, Andrew J. "Voyager: The Grand Tour of Big Science." In *From Engineering Science to Big Science: The NACA and NASA Collier Trophy Research Project Winners,* edited by Pamela E. Mack. NASA SP-4219. Washington, D.C., 1998.
Dick, Steven J., and LeRoy E. Doggett. *Sky with Ocean Joined: Proceedings of the Sesquicentennial Symposia of the U.S. Naval Observatory, December 5 and 8, 1980.* Washington, D.C.: U.S. Naval Observatory, 1983.
Douglas Aircraft Company. Missile and Space Systems Division, Atmospheric Sciences Branch. *Physical Properties of the Planet Mars.* Douglas Report SM-43634. August 1963.
Evans, D. C. *Physical Properties of the Planet Venus.* Douglas Report SM-41506. Missile & Space Systems Division, Douglas Aircraft. July 1962.
Frey, H., ed. *MEVTV Workshop on Early Tectonic and Volcanic Evolution of Mars.* Lunar and Planetary Institute Technical Report No. 89-04. Houston, Tex., October 5–7, 1988.

Hansen, James R. *The Atmosphere of Venus: Proceedings of a Conference Held at Goddard Institute for Space Studies, October 15–17, 1974.* New York: Goddard Institute for Space Studies, n.d. An informal version of the conference proceedings published in the *Journal of the Atmospheric Sciences,* June 1978.

Heacock, Raymond L. "The Voyager Spacecraft." James Watt International Gold Medal Lecture. *Proceedings of the Institution of Mechanical Engineers* 28 (1980): 194.

Hunt, Garry, ed. *Uranus and the Outer Planets: Proceedings of the IAU/RAS Colloquium No. 60.* Cambridge: Cambridge University Press, 1982.

Inquiry into Satellite and Missile Programs, Hearings before the Preparedness Investigating Subcommittee of the Committee on Armed Services. 85th Cong., 1st and 2d sess., pt. 1 and pt. 2.

Jet Propulsion Laboratory. *The Face of Mars.* Pasadena, Calif.: JPL, September 1975.

———. *The Many Faces of Mars.* JPL Technical Memorandum 33-654. Pasadena, Calif., December 1973.

———. Media Relations Office. Fact Sheets. Selected dates, 1997–98.

———. *Voyager at Neptune.* Pasadena, Calif.: JPL, California Institute of Technology, 1989.

———. *Voyager—Journey to the Outer Planets.* JPL-SP 43-39. Pasadena, Calif., n.d.

Kellog, William W., and Carl Sagan, preparers. *The Atmospheres of Mars and Venus: A Report by the Ad Hoc Panel on Planetary Atmospheres of the Space Science Board.* National Academy of Sciences–National Research Council publication 944. Washington, D.C.: National Academy of Sciences, 1962.

Leighton, Robert B., et al. *Mariner Mars 1964 Project Report: Television Experiment. Part I. Investigators' Report. Mariner IV Pictures of Mars.* Technical Report 32-884. Pasadena, Calif.: Jet Propulsion Laboratory, December 15, 1967.

Long, James E. "To the Outer Planets." *Astronautics & Aeronautics,* June 1969.

"Mars: An Informal Photographic Essay." *Transactions: American Geophysical Union* 57, no. 10 (October 1976); 1.

Mars Committee. "Minutes of a Meeting of the Mars Committee Held at the Headquarters of the National Geographic Society, Washington, D.C., March 29, 1954." N.d. Courtesy of William M. Sinton.

———. "Suggested Material for Program (Mars Conference–Lowell Observatory), October 22, 23, 1953. N.d. Courtesy of William M. Sinton.

———. "Suggested Topics for Discussion: Conference of the Mars Committee (to be Held at Lowell Observatory, October 22 and 23, 1953)." N.d. Courtesy of William M. Sinton.

Motions in Planetary Atmospheres: Proceedings of the Fourth Arizona Conference on Planetary Atmospheres. Kitt Peak National Observatory Special Contribution. Reprinted from *Earth and Extraterrestrial Science* 1 (December 1970): 171–84, and *Journal of Atmospheric Sciences* 27 (July 1970): 523–60.

NASA. Chronological Catalog of Reported Lunar Events. NASA TR R-277. Washington, D.C., July 1968.

———. *Mariner-Venus 1962: Final Project Report.* NASA SP-59. Washington, D.C., 1965.

———. *Models of Mars Atmosphere.* NASA SP-8010. Rev. Washington, D.C., December 1974.

———. *Models of the Venus Atmosphere.* NASA SP-8011. Washington, D.C., December 1968.

———. *Pioneer Odyssey: Encounter with a Giant.* NASA SP-1-349. Washington, D.C., 1974.

———. *The Planet Saturn*. NASA SP-8091. Washington, D.C., June 1972.

———. *Surface Models of Mars*. NASA SP-8020. Rev. Washington, D.C., September 1975.

National Academy of Sciences–National Research Council. *Astronomy and Astrophysics for the 1980's*. Vol. 1, *Report of the Astronomy Survey Committee*. Final prepublication draft. Washington, D.C.: NAS-NRC, 1982. Author's papers. This is the "Field report."

———. *Ground-Based Astronomy a Ten Year Program: A Report Prepared by the Panel on Astronomical Facilities for the Committee on Science and Public Policy of the National Academy of Sciences*. Washington, D.C.: NAS-NRC, 1964. This is the "Whitford Report."

———. *Space Research: Directions for the Future*. Washington, D.C.: NAS-NRC, 1966.

National Academy of Sciences–Space Science Board–National Research Council. *Priorities for Space Research*. Washington, D.C.: NAS-NRC, 1971.

———. *A Review of Space Research*. Publication 1079. Washington, D.C.: National Academy of Sciences, 1962.

National Research Council. Space Studies Board. *A Science Strategy for the Exploration of Europa*. Washington, D.C.: National Academy Press, 1999.

———. *Preventing the Forward Contamination of Europa*. Washington, D.C.: National Academy Press, 2000.

———. Task Group on Planetary and Lunar Exploration. *Space Science in the Twenty-First Century: Imperatives for the Decades 1995 to 2015*. Washington, D.C.: National Academy Press, 1988.

Peebles, Curtis. "The Original Voyager: A Mission Not Flown." *Journal of the British Interplanetary Society* 35 (1982).

Pioneer 10 Mission: Jupiter Encounter. Reprinted from the *Journal of Geophysical Research* 79, no. 10 (September 1, 1974): 3487.

"Pioneer Saturn." Reprinted from the *Journal of Geophysical Research* 85, no. A11 (November 1, 1980): 5651.

"The Planet Venus: Past, Present, and Future." *Proceedings of the American Philosophical Society* 113, no. 3 (June 16, 1969): 197.

"Problems in Planetary Physics: A Symposium Held in Connection with the Dedication of a Planetary Research Center at the Lowell Observatory on May 3, 1963." *Lowell Observatory Bulletin* 6, no. 9, serial no. 128 (n.d., but introductory material is dated June 3, 1965), 179.

Sinton, William M., "Distribution of Temperatures and Spectra of Venus and Other Planets." Ph.D. diss., Johns Hopkins University, 1953.

Sinton, William M., and John Strong. "Observations of the Infrared Emission of Planets and Determination of Their Temperatures." Progress report for the Office of Naval Research. Baltimore: Johns Hopkins University, Laboratory of Astrophysics and Physical Meteorology, April 15, 1960.

Slipher, Earl C., and Gerard de Vaucouleurs. "International Mars Committee (IMC): Circular to members on programs and projects, 1958–59." Letter and tentative, provisional list of programs and projects for members to fill out. N.d. Courtesy of William M. Sinton.

Slipher, Earl C., Gerard de Vaucouleurs, and A. G. Wilson. "Report on the Conference of the Mars Committee (Held at Lowell Observatory, October 22, 23, 1953). N.d., but not later than November 6, 1953. Courtesy of William M. Sinton.

The Solar System. A Scientific American book reprinted from *Scientific American,* September 1975. San Francisco: W. H. Freeman, 1975.

Special reprint of articles about the Pioneer 10 Venus encounter from *Science* 183, no. 4131 (March 29, 1974): 1289–1321.

Staff Report of the Select Committee on Astronautics and Space Exploration. *The Next Ten Years in Space.* Washington, D.C.: GPO, 1959.

Strong, John, and William M. Sinton. "Radiometry of Mars." Report on July 1954 observations. N.d., but not later than October 18, 1954. Courtesy of William M. Sinton.

Struve, Otto. "The Birth of McDonald Observatory." *Sky & Telescope* 24 (1962): 316.

TRW Space Log 9, no. 4 (Winter 1969–70).

Waff, Craig B. "The Road to the Deep Space Network." *Spectrum,* April 1993, 50.

Wilson Albert G. "For Mars Committee Secretaries, to William Sinton containing a request for details of work on Mars during the 1954 opposition." October 4, 1954. Courtesy of William M. Sinton.

———. "For the Secretaries of the Mars Committee, to William Sinton mainly concerned with publicity and news releases about Mars." April 5, 1954. Courtesy of William M. Sinton.

COMPACT DISCS

Murmurs of Earth. Vox Records, n.d.

Symphonies of the Planets. Vols. 1–4. Los Angeles: Delta Music, 1992.

SPECIAL JOURNAL ISSUES

Icarus 53 (1983): 163–387. Voyager 1 and 2 at Saturn.

Icarus 54 (1983): 159–360. Voyager 1 and 2 at Saturn.

Journal of Geophysical Research 85 (1981): 8123–841. Voyagers 1 and 2 at Jupiter.

Journal of Geophysical Research 88 (1983): 8625–9018. Voyagers 1 and 2 at Saturn.

Journal of Geophysical Research 92 (1987): 14,873–15,375. Voyager 2 at Uranus.

Journal of Geophysical Research 96, supplement (1991): 18,903–19,268. Voyager 2 at Neptune. This supplement is often bound separately and placed on library shelves at the end of volume 96, a location that is not consistent with its page numbers

Journal of Geophysical Research Letters 7 (1980): 1–68. Voyager 1 at Jupiter.

Nature 280 (1979): 725–806. Voyager 1 at Jupiter.

Nature 292 (1981): 675–755. Voyager 1 at Saturn.

Science 204 (1979): 945–1008. Voyager 1 at Jupiter.

Science 206 (1979): 925–96. Voyager 2 at Jupiter.

Science 212 (1981): 159–243. Voyager 1 at Saturn.

Science 215 (1984): 499–595. Voyager 2 at Saturn.

Science 233 (1986): 39–109. Voyager 2 at Uranus.

Science 246 (1989): 1417–1501. Voyager 2 at Neptune.

INDEX